Standard Grade | Credit

Biology

Credit Level 2004

Credit Level 2005

Credit Level 2006

Credit Level 2007

Credit Level 2008

Leckie × Leckie

First exam published in 2004.
Published by Leckie & Leckie Ltd, 3rd Floor, 4 Queen Street, Edinburgh EH2 1JE
tel: 0131 220 6831 fax: 0131 225 9987 enquiries@leckieandleckie.co.uk www.leckieandleckie.co.uk

ISBN 978-1-84372-621-0

A CIP Catalogue record for this book is available from the British Library.

Leckie & Leckie is a division of Huveaux plc.

Leckie & Leckie is grateful to the copyright holders, as credited at the back of the book, for permission to use their material. Every effort has been made to trace the copyright holders and to obtain their permission for the use of copyright material. Leckie & Leckie will gladly receive information enabling them to rectify any error or omission in subsequent editions.

BLANK PAGE

C

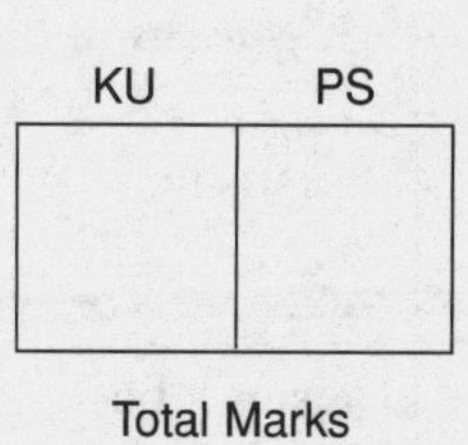

0300/402

NATIONAL QUALIFICATIONS 2004

WEDNESDAY, 19 MAY
10.50 AM – 12.20 PM

BIOLOGY
STANDARD GRADE
Credit Level

Fill in these boxes and read what is printed below.

Full name of centre

Town

Forename(s)

Surname

Date of birth
Day Month Year

Scottish candidate number

Number of seat

1 All questions should be attempted.

2 The questions may be answered in any order but all answers are to be written in the spaces provided in this answer book, and must be written clearly and legibly in ink.

3 Rough work, if any should be necessary, as well as the fair copy, is to be written in this book. Additional spaces for answers and for rough work will be found at the end of the book. Rough work should be scored through when the fair copy has been written.

4 Before leaving the examination room you must give this book to the invigilator. If you do not, you may lose all the marks for this paper.

LIB 0300/402 6/26620

DO NOT WRITE IN THIS MARGIN

Marks | KU | PS

1. The table contains information about five species of bat.

Species	*Wingspan* (cm)	*Roosting place*	*Flight*
Pipistrelle bat	19–25	Trees and buildings	Fast and erratic
Leisler's bat	25–33	Trees and buildings	Fast and straight
Lesser Horseshoe bat	19–25	Buildings only	Fast and agile
Bechstein's bat	25–33	Trees and buildings	Slow and fluttering
Daubenton's bat	19–25	Trees and buildings	Fast and straight

Use the information from the table to complete the boxes of the paired statement key below.

1 Wingspan 19–25 cm .... [ ]

Wingspan 25–33 cm .... go to 3

2 Roosts in [ ] .... Lesser Horseshoe bat

Roosts in trees and buildings .... go to 4

3 Slow and fluttering flight .... [ ]

Fast and straight flight .... [ ]

4 [ ] .... Pipistrelle bat

[ ] .... Daubenton's bat

3

DO NOT WRITE IN THIS MARGIN

Marks KU PS

2. The diagram shows part of a food web from the sea.

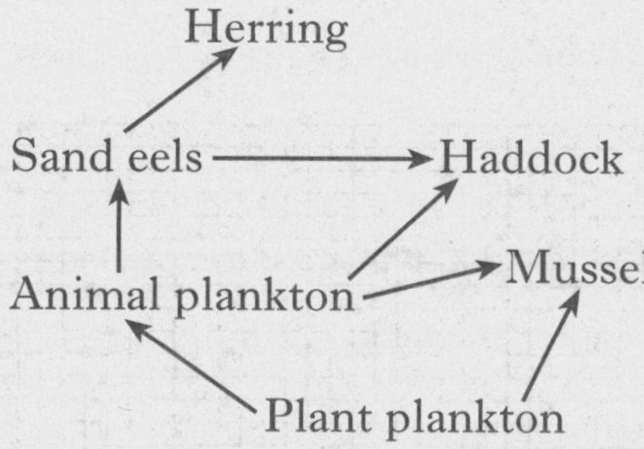

Herring

Sand eels → Haddock

Animal plankton → Mussel

Plant plankton

(*a*) Cod are fish that feed on young herring and sand eels.

Complete the food web to show the relationship of the cod to the other organisms in the food web. **1**

(*b*) Over fishing has led to a decrease in the numbers of haddock in the food web.

(i) Explain why the population of animal plankton may **increase** if the haddock numbers are reduced.

____

____ **1**

(ii) Explain why the population of animal plankton may **stay the same** if the haddock numbers are reduced.

____

____ **1**

(*c*) Underline **one** word in each pair to make the following sentences correct.

{Producers / Consumers} have the greatest biomass in a food chain. At each stage in the food chain, the biomass is {greater / smaller} than the stage before.

The reason for this is that energy is {gained / lost} from each stage. **2**

[Turn over

DO NOT WRITE IN THIS MARGIN

Marks | KU | PS

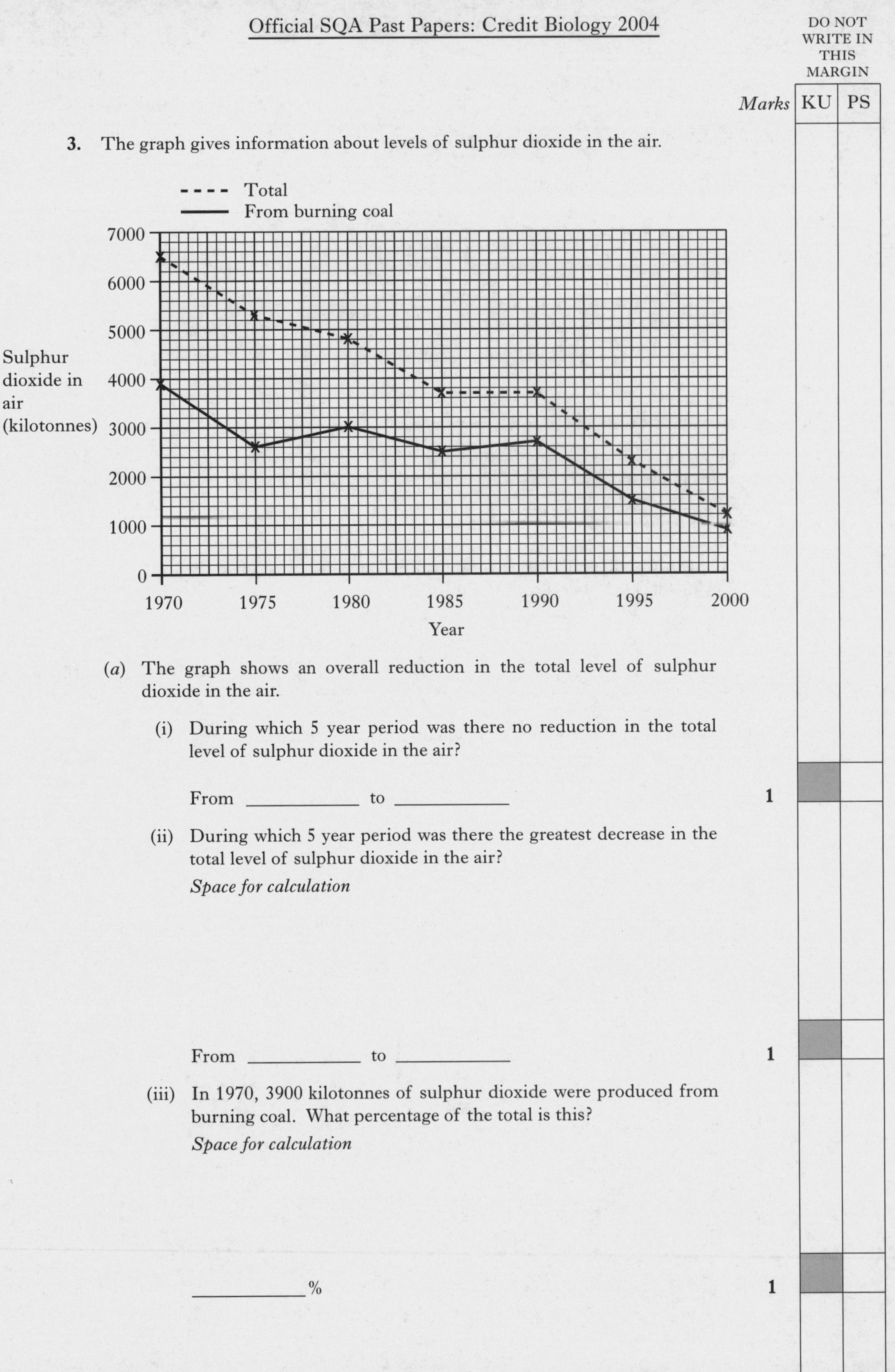

3. The graph gives information about levels of sulphur dioxide in the air.

(*a*) The graph shows an overall reduction in the total level of sulphur dioxide in the air.

(i) During which 5 year period was there no reduction in the total level of sulphur dioxide in the air?

From ____________ to ____________ **1**

(ii) During which 5 year period was there the greatest decrease in the total level of sulphur dioxide in the air?

Space for calculation

From ____________ to ____________ **1**

(iii) In 1970, 3900 kilotonnes of sulphur dioxide were produced from burning coal. What percentage of the total is this?

Space for calculation

____________ % **1**

DO NOT WRITE IN THIS MARGIN

Marks | KU | PS

3. (*a*) (continued)

(iv) Underline the correct option in the following sentence.

Between 1970 and 2000 the proportion of sulphur dioxide in the air which came from burning coal {increased / decreased / stayed the same}. **1**

(*b*) Sulphur dioxide levels in the air could be reduced by switching from coal-fired power stations to nuclear power stations.
Give one **disadvantage** of using nuclear power.

____

____ **1**

[Turn over

DO NOT WRITE IN THIS MARGIN

Marks KU PS

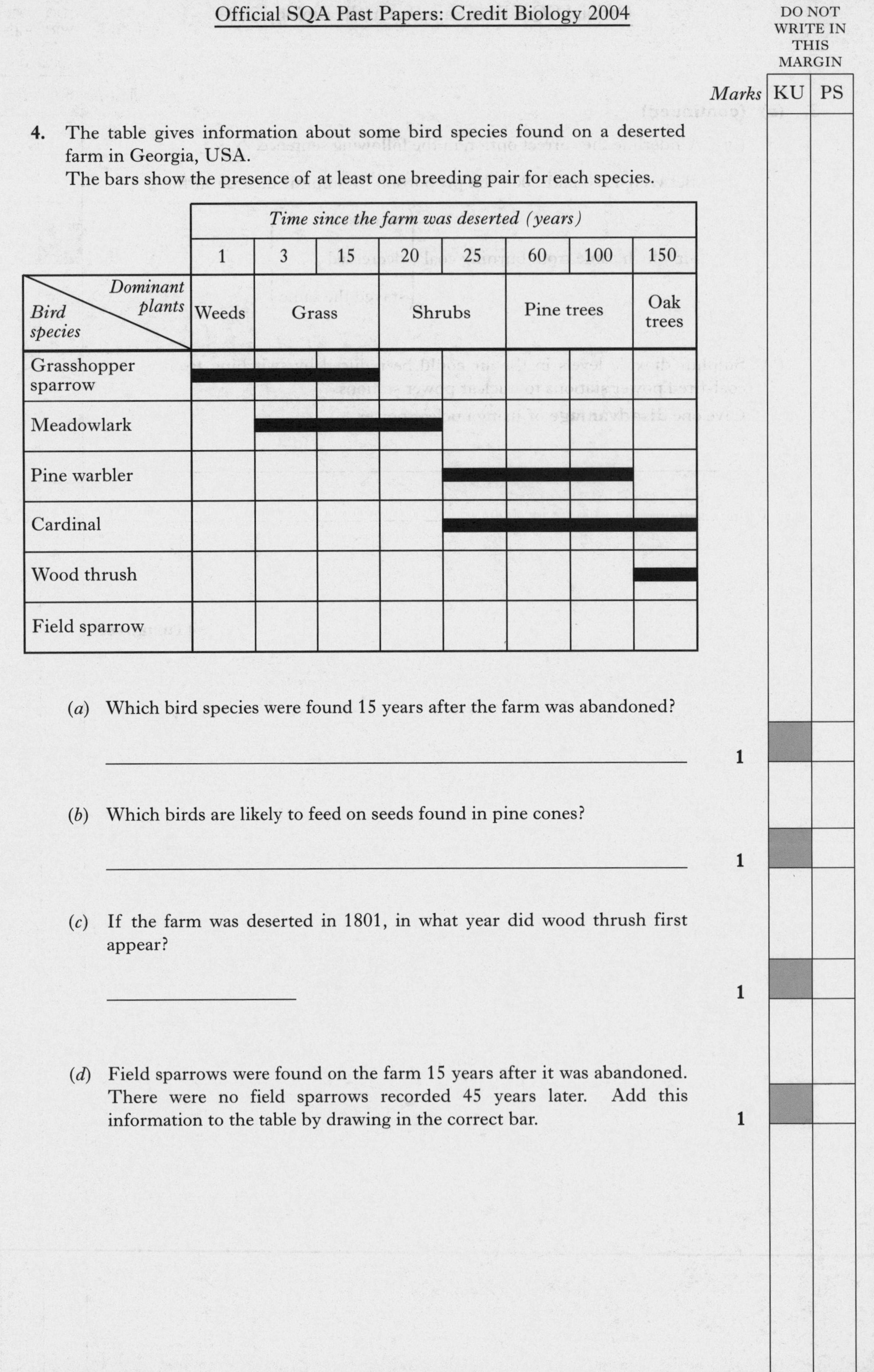

4. The table gives information about some bird species found on a deserted farm in Georgia, USA.
The bars show the presence of at least one breeding pair for each species.

	Time since the farm was deserted (years)							
	1	3	15	20	25	60	100	150
Bird species / *Dominant plants*	Weeds	Grass		Shrubs		Pine trees		Oak trees
Grasshopper sparrow	■	■	■					
Meadowlark		■	■	■				
Pine warbler					■	■	■	
Cardinal					■	■	■	■
Wood thrush								■
Field sparrow								

(*a*) Which bird species were found 15 years after the farm was abandoned?

______________________________ **1**

(*b*) Which birds are likely to feed on seeds found in pine cones?

______________________________ **1**

(*c*) If the farm was deserted in 1801, in what year did wood thrush first appear?

______________ **1**

(*d*) Field sparrows were found on the farm 15 years after it was abandoned. There were no field sparrows recorded 45 years later. Add this information to the table by drawing in the correct bar. **1**

DO NOT WRITE IN THIS MARGIN

Marks | KU | PS

5. (a) The following statements refer to the stages that occur after pollination.

A Fertilisation takes place.
B A pollen tube grows out from a pollen grain.
C The ovule forms a seed and the ovary forms a fruit.
D The pollen tube grows down through the stigma.
E The male gamete moves towards the ovule.
F The pollen tube grows through the ovary wall.

Use the letters of the statements to complete the sequence of stages.

B → ☐ → ☐ → E → ☐ → ☐ **2**

(b) Plants can reproduce by sexual and asexual means.

Draw lines to link each method of reproduction with the advantages that each method provides.

Method of reproduction	*Advantage*
	Offspring obtains food and water from parent
Sexual	Seeds are dispersed
Asexual	Greater variation among offspring
	Pollination is not required

2

(c) The plants in a clone have been produced by asexual reproduction.

Give **one** other piece of information about the members of a clone.

____

____ **1**

[Turn over

DO NOT WRITE IN THIS MARGIN

Marks | KU | PS

6. (*a*) Complete the following table about the three major food groups.

Type of food	*Chemical elements present*	*Example of digestive enzyme*	*Product(s) of digestion*
Carbohydrates	1 carbon 2 hydrogen 3 oxygen		
Fats	1 2 3	lipase	1 fatty acids 2 glycerol
	1 carbon 2 hydrogen 3 oxygen 4 nitrogen		

3

(*b*) The villi which line the small intestine each contain a lacteal and blood capillaries.

Give a brief description of the function of each of these structures.

Lacteal ______________________________

______________________________ **1**

Blood capillaries ______________________________

______________________________ **1**

DO NOT WRITE IN THIS MARGIN

Marks KU PS

7. The diagram shows a developing human fetus.

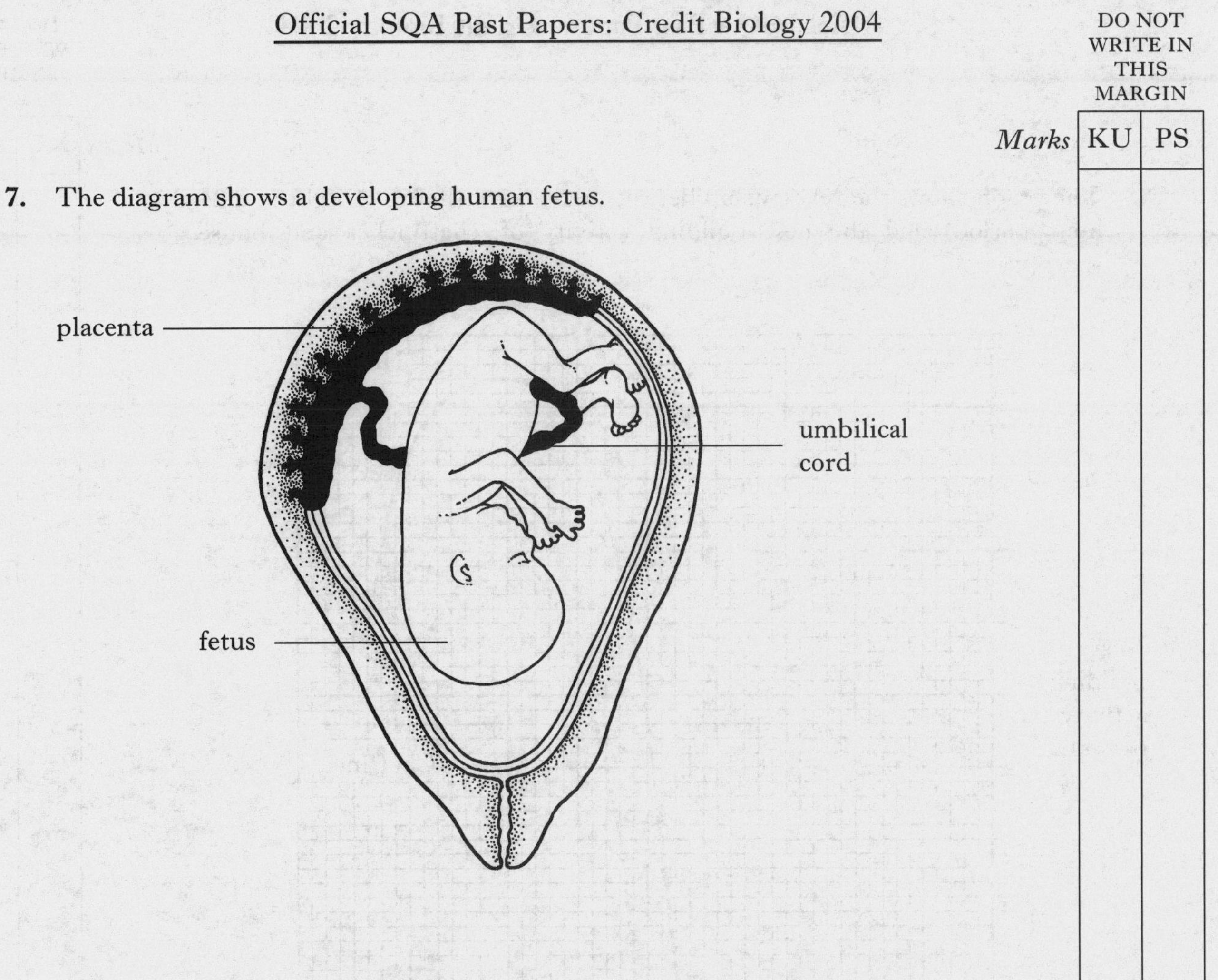

Complete the following table by putting a tick in the correct column to indicate the main direction of exchange for each of the following materials.

The first one has been done for you.

Material	*Direction of exchange*		
	Mother to fetus	*Fetus to mother*	*No exchange*
glucose	✓		
amino acids			
blood			
oxygen			
urea			
carbon dioxide			

2

[Turn over

DO NOT WRITE IN THIS MARGIN

Marks | KU | PS

8. The graph shows the relationship between daylength (hours of light in a 24 hour period) and the nest building activity of chaffinches and house sparrows.

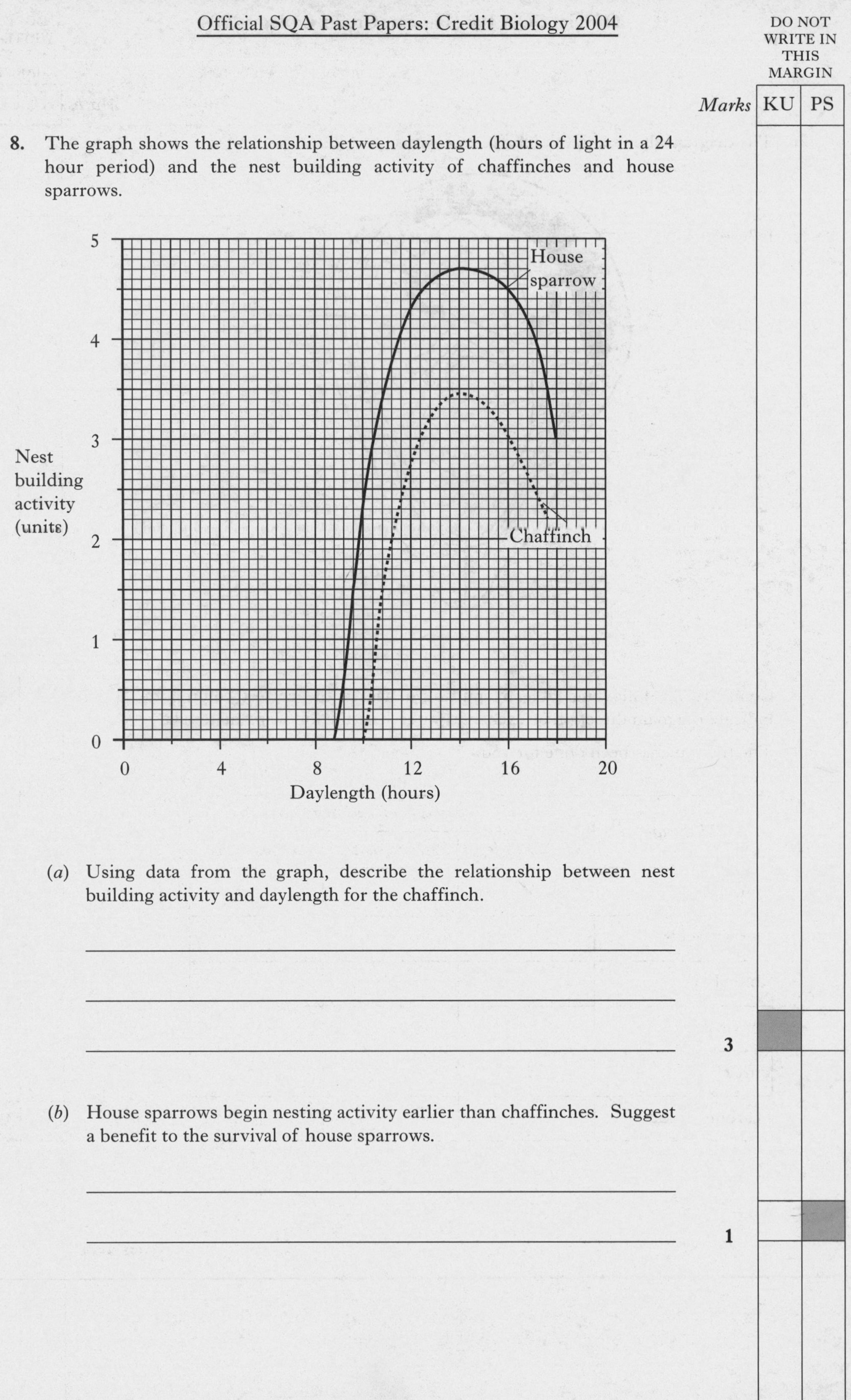

(*a*) Using data from the graph, describe the relationship between nest building activity and daylength for the chaffinch.

____

____

____ **3**

(*b*) House sparrows begin nesting activity earlier than chaffinches. Suggest a benefit to the survival of house sparrows.

____

____ **1**

DO NOT WRITE IN THIS MARGIN

Marks KU PS

9. Plant cells and animal cells were left in water or 10% sucrose solution for 10 minutes. The cells were then examined under the microscope. The appearance of three individual cells is shown below.

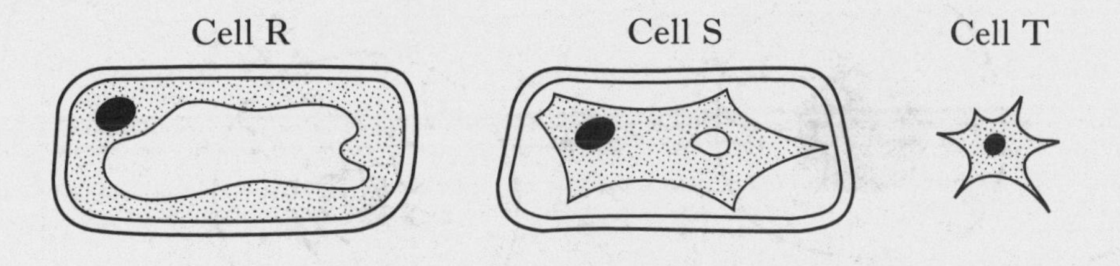

(*a*) Which **two** of the cells had been placed in 10% sucrose solution?

Cell ________ and Cell ________ **1**

(*b*) The change in the cells was caused by the movement of water into or out of the cells.

What is the name of this process?

______________________________ **1**

(*c*) With reference to the cells placed in water, what is meant by the term "concentration gradient"?

____

____ **1**

[Turn over

DO NOT WRITE IN THIS MARGIN

Marks KU PS

10. The diagram represents some of the stages of cell division.

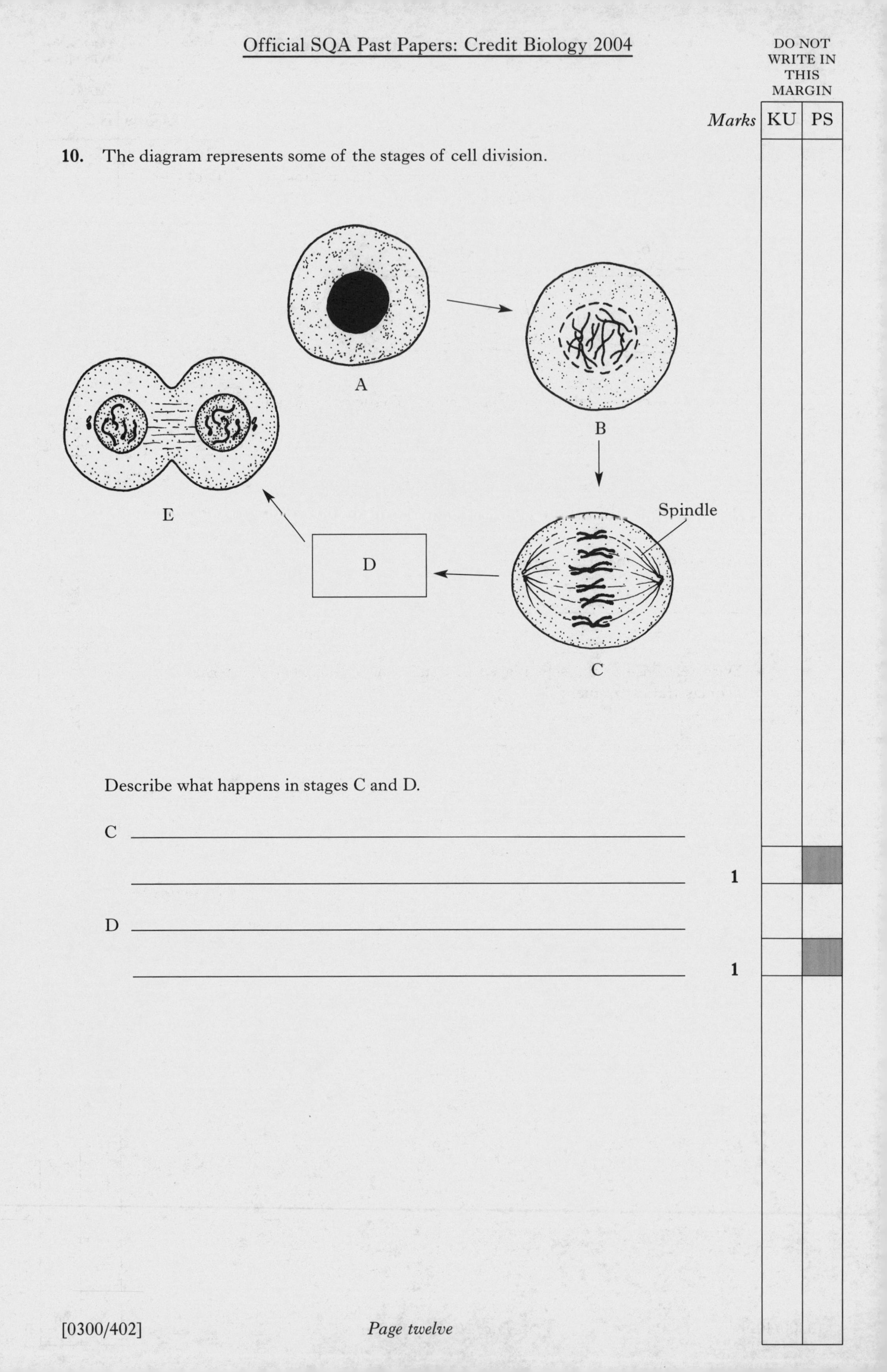

Describe what happens in stages C and D.

C ____

____ **1**

D ____

____ **1**

DO NOT WRITE IN THIS MARGIN

Marks | KU | PS

11. The activity of the enzymes lipase and catalase was investigated.

Three test tubes were set up.

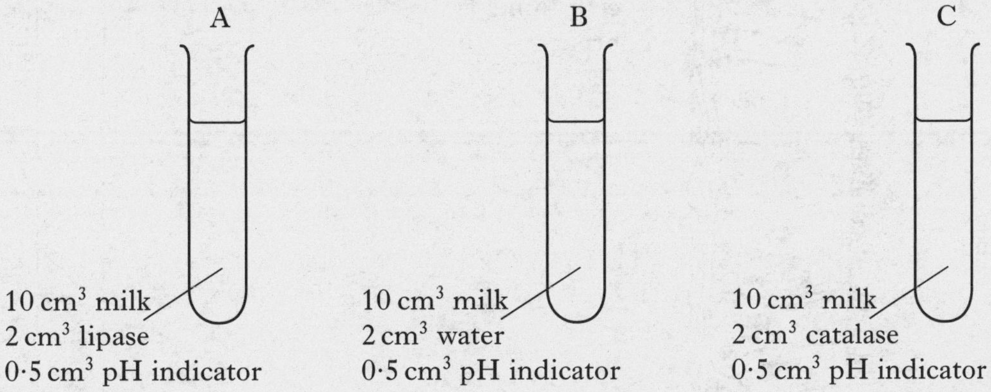

A

10 cm^3 milk
2 cm^3 lipase
0·5 cm^3 pH indicator

B

10 cm^3 milk
2 cm^3 water
0·5 cm^3 pH indicator

C

10 cm^3 milk
2 cm^3 catalase
0·5 cm^3 pH indicator

The colour of the pH indicator was noted at the start and after 20 minutes.

The results are shown in the table below.

	Colour of pH indicator	
Test tube	*At start*	*After 20 minutes*
A	green	orange
B	green	green
C	green	green

(*a*) In tube A, the pH indicator colour change was due to the production of fatty acids as the lipase reacted with the fat in the milk.

Explain why there was no change in tube C.

____

______________________________ **1**

(*b*) What term is used to describe tube B which contained water instead of an enzyme?

______________ **1**

(*c*) Name **two** variables, not already shown, which would have to be kept the same when this investigation was set up.

1 ______________________________

2 ______________________________ **2**

DO NOT WRITE IN THIS MARGIN

Marks | KU | PS

12. The diagram represents part of the breathing system in humans.

cells lining a bronchus

mucus secreting cell

mucus

cilia

(*a*) Describe how the mucus and cilia help to protect the lungs from damage and infection.

Mucus ____

____

Cilia ____

____ **2**

(*b*) Which of the following are involved in **breathing out** during deep breathing in humans?

1 Diaphragm contracts
2 Diaphragm relaxes
3 Muscles between the ribs contract
4 Muscles between the ribs relax
5 Rib cage moves up and out
6 Rib cage moves down and in

Tick the correct box.

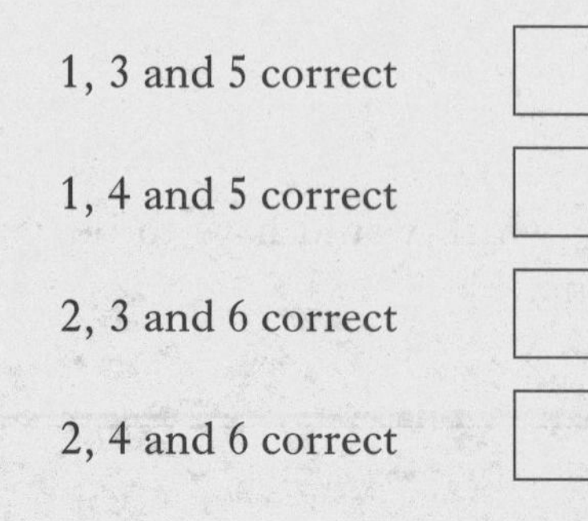

1, 3 and 5 correct ☐

1, 4 and 5 correct ☐

2, 3 and 6 correct ☐

2, 4 and 6 correct ☐ **1**

DO NOT WRITE IN THIS MARGIN

Marks	KU	PS

12. (continued)

(*c*) The pie charts below show the composition of fresh air and breathed air.

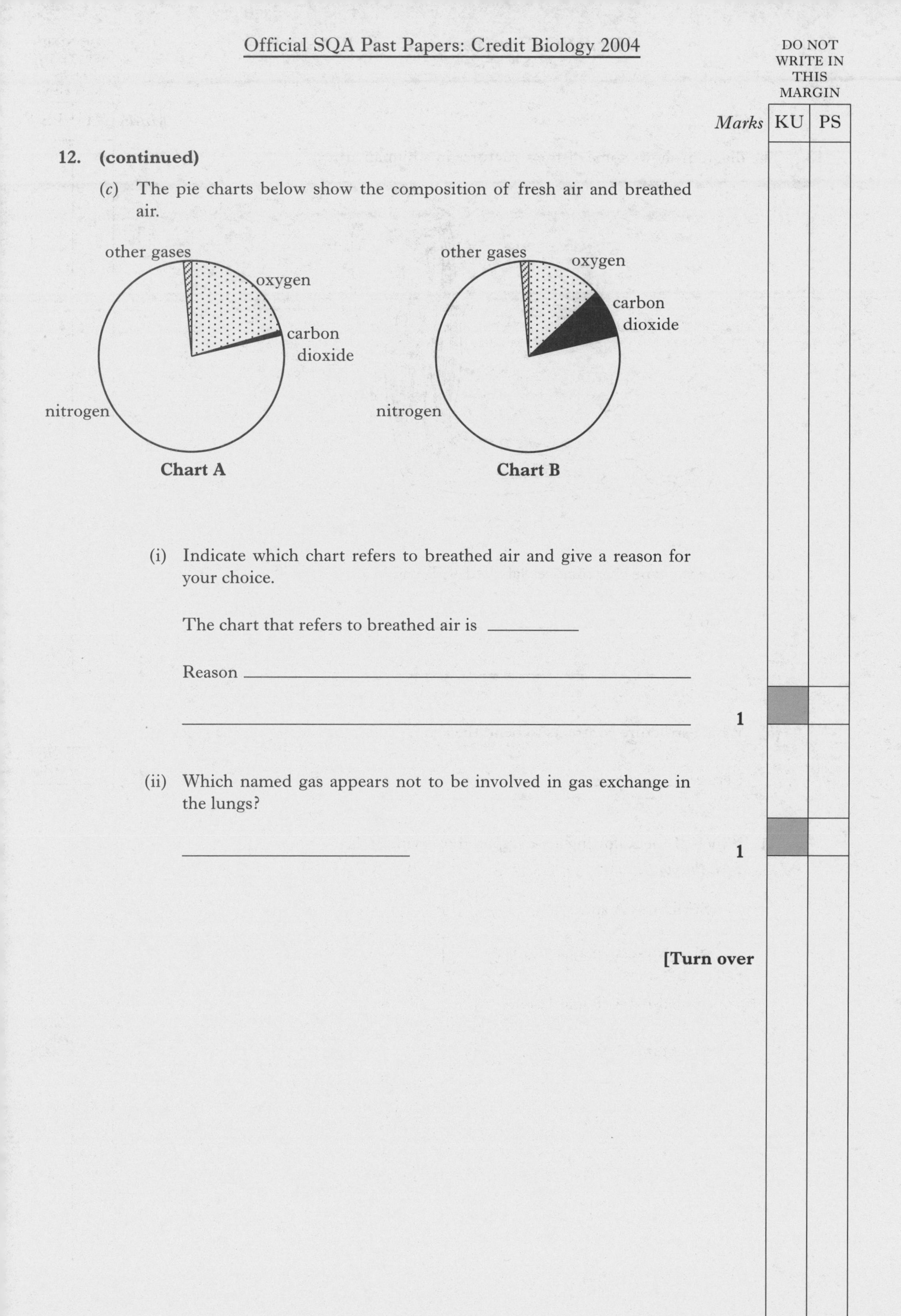

(i) Indicate which chart refers to breathed air and give a reason for your choice.

The chart that refers to breathed air is ________

Reason ____

____ **1**

(ii) Which named gas appears not to be involved in gas exchange in the lungs?

________________________ **1**

[Turn over

DO NOT WRITE IN THIS MARGIN

Marks KU PS

13. The diagram shows some of the structures in a human arm.

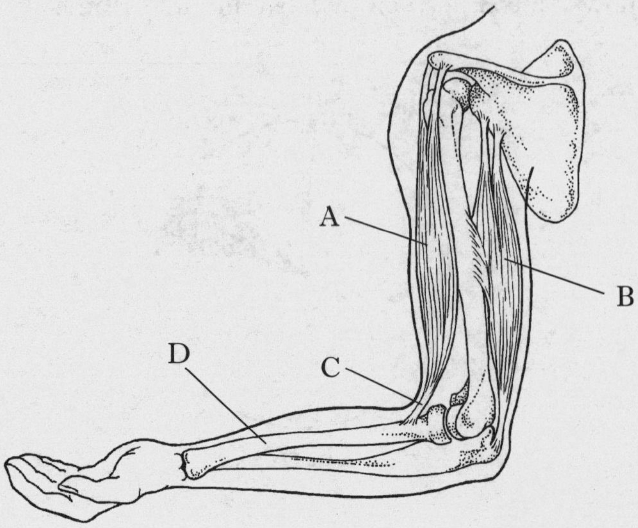

(*a*) Name the type of structures labelled A, B and C.

A and B ______________________________

C ______________________________ **1**

(*b*) Which structure contracts to bend the arm?

Letter ________ **1**

(*c*) Which of the following are composed of living cells?
Tick the correct box.

Structures A and B only ☐

Structures A, B and C only ☐

Structures A, B and D only ☐

Structures A, B, C and D ☐ **1**

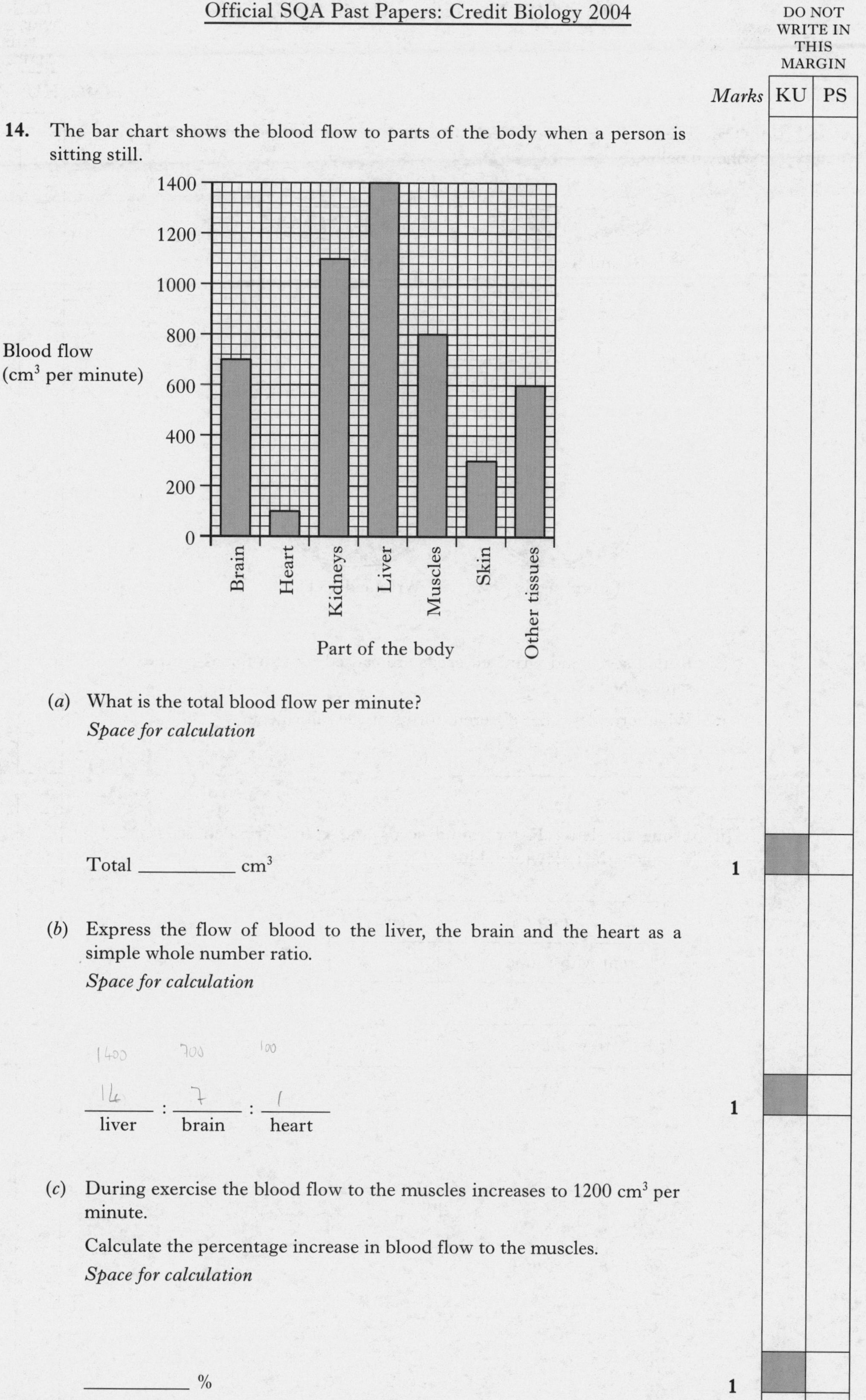

DO NOT WRITE IN THIS MARGIN

Marks	KU	PS

14. The bar chart shows the blood flow to parts of the body when a person is sitting still.

(*a*) What is the total blood flow per minute?
Space for calculation

Total __________ cm^3 **1**

(*b*) Express the flow of blood to the liver, the brain and the heart as a simple whole number ratio.
Space for calculation

1400 700 100

14 : 7 : 1
liver : brain : heart **1**

(*c*) During exercise the blood flow to the muscles increases to 1200 cm^3 per minute.

Calculate the percentage increase in blood flow to the muscles.
Space for calculation

__________ % **1**

[Turn over

DO NOT WRITE IN THIS MARGIN

Marks | KU | PS

15. (*a*) True breeding pea plants were bred to produce two generations, as shown below.

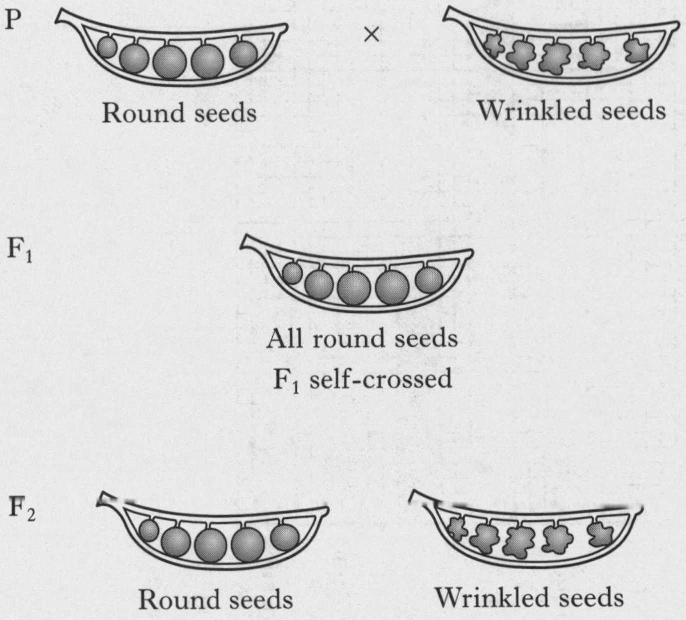

P — Round seeds × Wrinkled seeds

F_1 — All round seeds

F_1 self-crossed

F_2 — Round seeds, Wrinkled seeds

(i) Round seeds and wrinkled seeds are caused by two forms of the same gene.

What term describes different forms of the same gene?

______________________________ **1**

(ii) Using the letter **R** for round seeds and **r** for wrinkled seeds, complete the following table.

Plant	*Genotype*
Parent with round seeds	
All F_1	
F_2 with wrinkled seeds	

2

DO NOT WRITE IN THIS MARGIN

Marks KU PS

15. (*a*) (continued)

(iii) The seeds from the F_2 were counted and the results are shown in the bar chart.

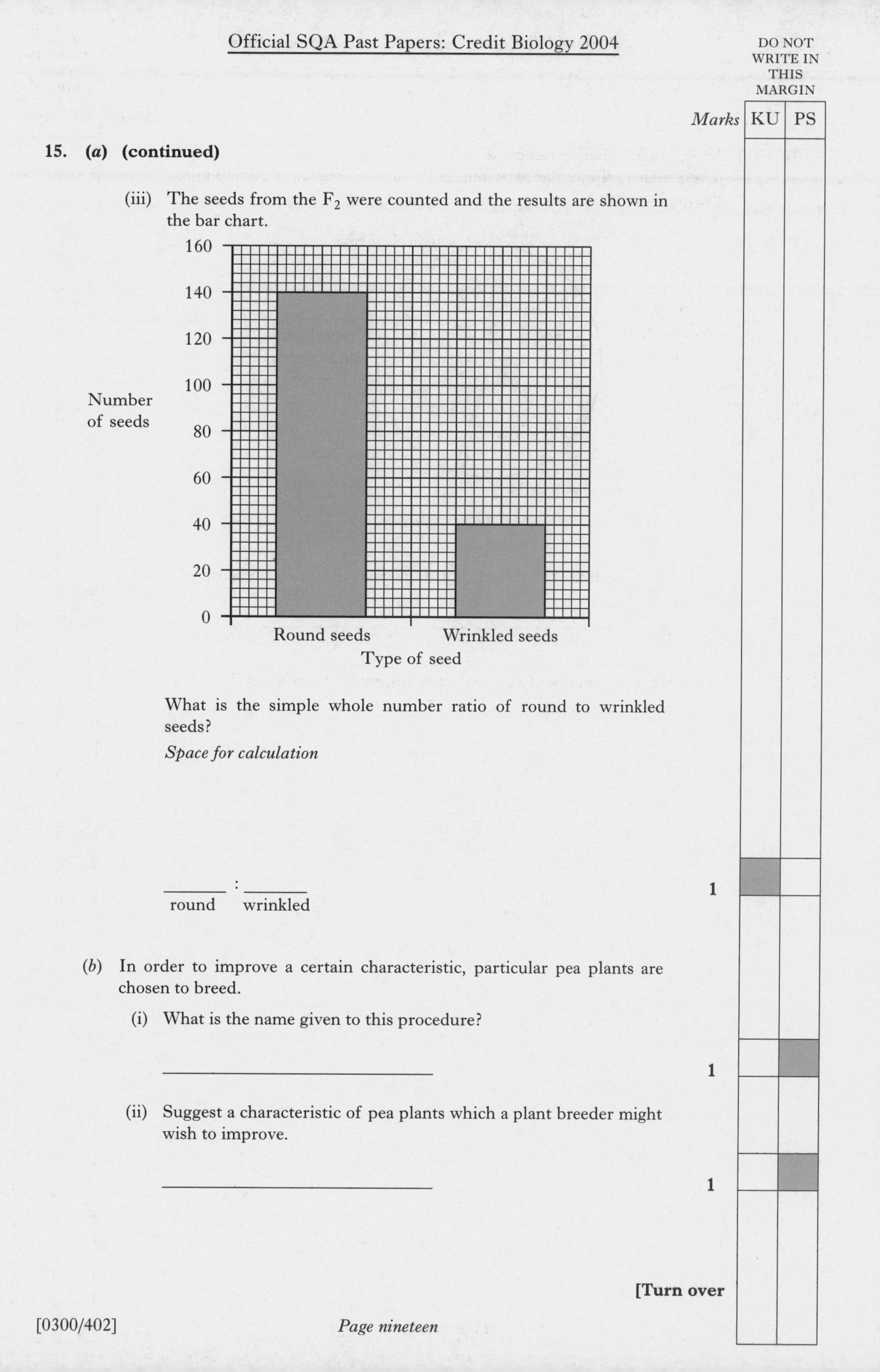

What is the simple whole number ratio of round to wrinkled seeds?

Space for calculation

______ : ______

round   wrinkled

1

(*b*) In order to improve a certain characteristic, particular pea plants are chosen to breed.

(i) What is the name given to this procedure?

____

1

(ii) Suggest a characteristic of pea plants which a plant breeder might wish to improve.

____

1

[Turn over

DO NOT WRITE IN THIS MARGIN

Marks | KU | PS

16. (*a*) Exposure to radiation can cause mutation.
The pie chart shows the contribution of various sources of radiation to the total exposure.

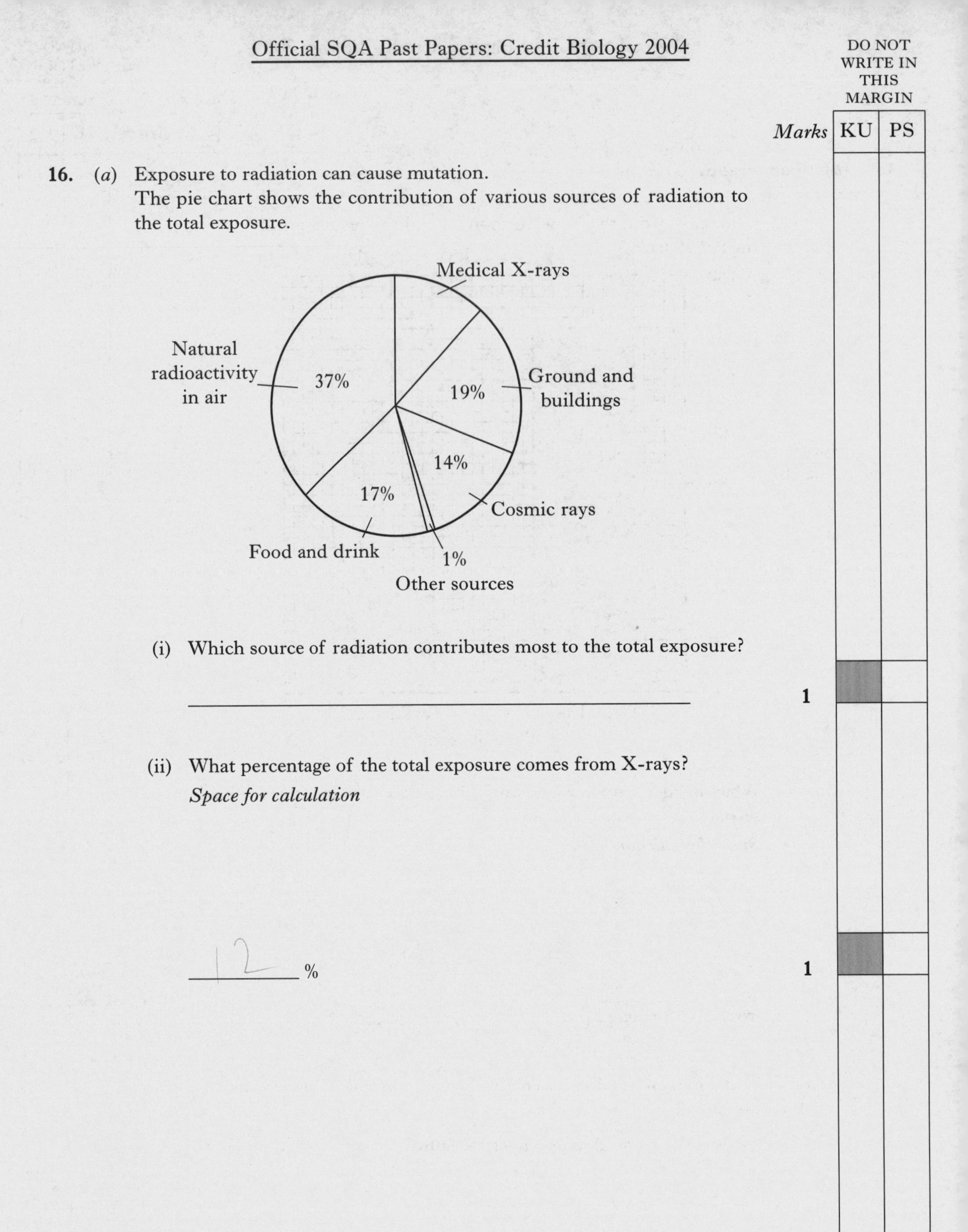

(i) Which source of radiation contributes most to the total exposure?

________________________________ **1**

(ii) What percentage of the total exposure comes from X-rays?
Space for calculation

12 % **1**

Marks | KU | PS

16. (continued)

(*b*) The table shows the occurrence of chromosome mutations in *Drosophila* fruit flies when exposed to different doses of radiation.

Dosage of X-rays (millisieverts)	*Chromosome mutations* (%)
1000	1·0
2000	1·9
2500	2·6
3000	3·1
4000	4·2
4500	4·6
5000	5·3

(i) On the grid below, complete the y-axis and plot a line graph of the results.

(An additional grid, if needed, will be found on page 27.)

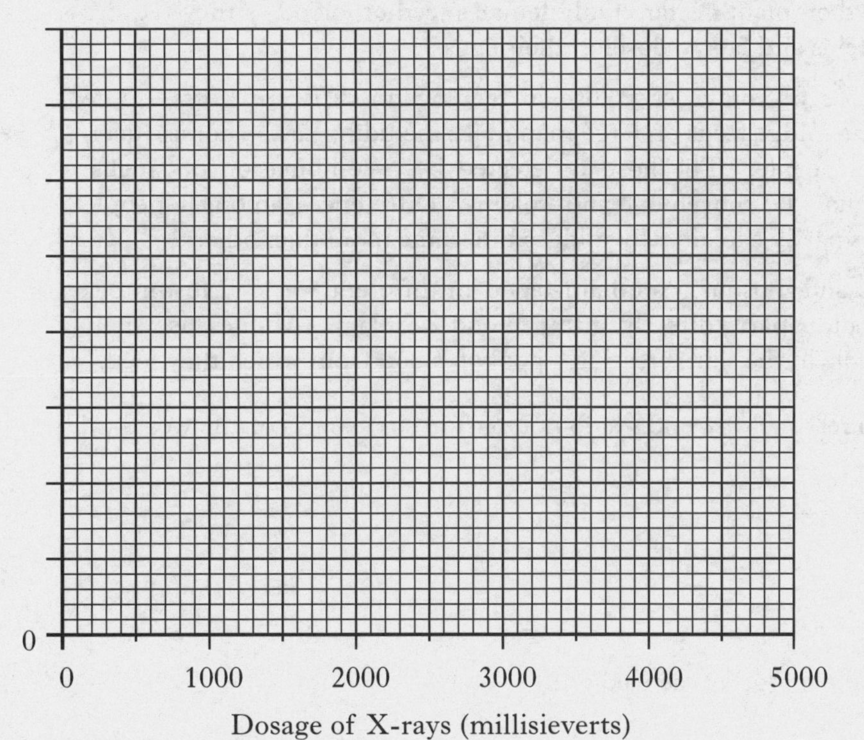

2

(ii) What is the relationship between the dosage of X-rays and chromosome mutations?

____

____ **1**

DO NOT WRITE IN THIS MARGIN

Marks | KU | PS

17. Read the following passage and answer the questions based on it.

Midges are blood-feeding insects similar to mosquitoes but much smaller. There are about 40 species found in Scotland. One species, *Culicoides impunctatus* (or the 'Highland biting midge'), is best known for biting people.

Midges inhabit all areas of the world except the poles, New Zealand and Patagonia (where the habitat is too dry). In the UK, their numbers are greatest in parts of Western Scotland and the Highlands where they thrive in the damp acidic soil. The colder winters on the East coast often result in frozen soil which kills the overwintering larvae.

In late summer, eggs are laid on the soil surface. They hatch into thread-like larvae which live a few centimetres below the soil surface and feed on decaying plant and animal matter. Flying adults begin to emerge the following May, triggered by lengthening days and warmer temperatures. These adults lay eggs that develop quickly to give a second emergence of adult midges in August.

Midges are unable to fly in very wet or windy weather or in temperatures below 7 °C. In such poor conditions they might only survive a few days. In more favourable conditions they may live for two weeks.

All midges feed on plant nectar. Only females feed on blood as they require the blood proteins and fats to develop their eggs.

Different species of midge specialise in feeding on different hosts. The Highland biting midge feeds on large mammals, including cattle, horses, deer, and of course, people. The host is detected by a combination of smells, heat, carbon dioxide, movement and colour. Differences amongst people in these factors partly explain why midges bite some more than others.

Biting midges and mosquitoes obtain blood in different ways. Mosquitoes insert their mouth-parts directly into a blood capillary. Midges use their jaws to cut a hole in the skin, creating a pool of blood from which they feed.

(*a*) Give **two** reasons why midges do not exist in some parts of the world.

1 ____

2 ____ **1**

(*b*) What do the midge larvae feed on?

____ **1**

(*c*) What is the maximum time an adult midge can live for?

____ **1**

DO NOT WRITE IN THIS MARGIN

	Marks	KU	PS

17. (continued)

(*d*) Name **two** types of food that an adult female midge might eat.

1 ____________________________________

2 ____________________________________ **1**

(*e*) Suggest why the Highland biting midge might not bite mice.

____

____________________________________ **1**

(*f*) Describe **two** differences, mentioned in the passage, between midges and mosquitoes.

1 ____________________________________

2 ____________________________________ **1**

[Turn over

DO NOT WRITE IN THIS MARGIN

Marks | KU | PS

18. An investigation into fermentation was carried out at 20 °C.

A deflated balloon was attached to the top of each tube at the start. The appearance of the balloons after several hours is shown below.

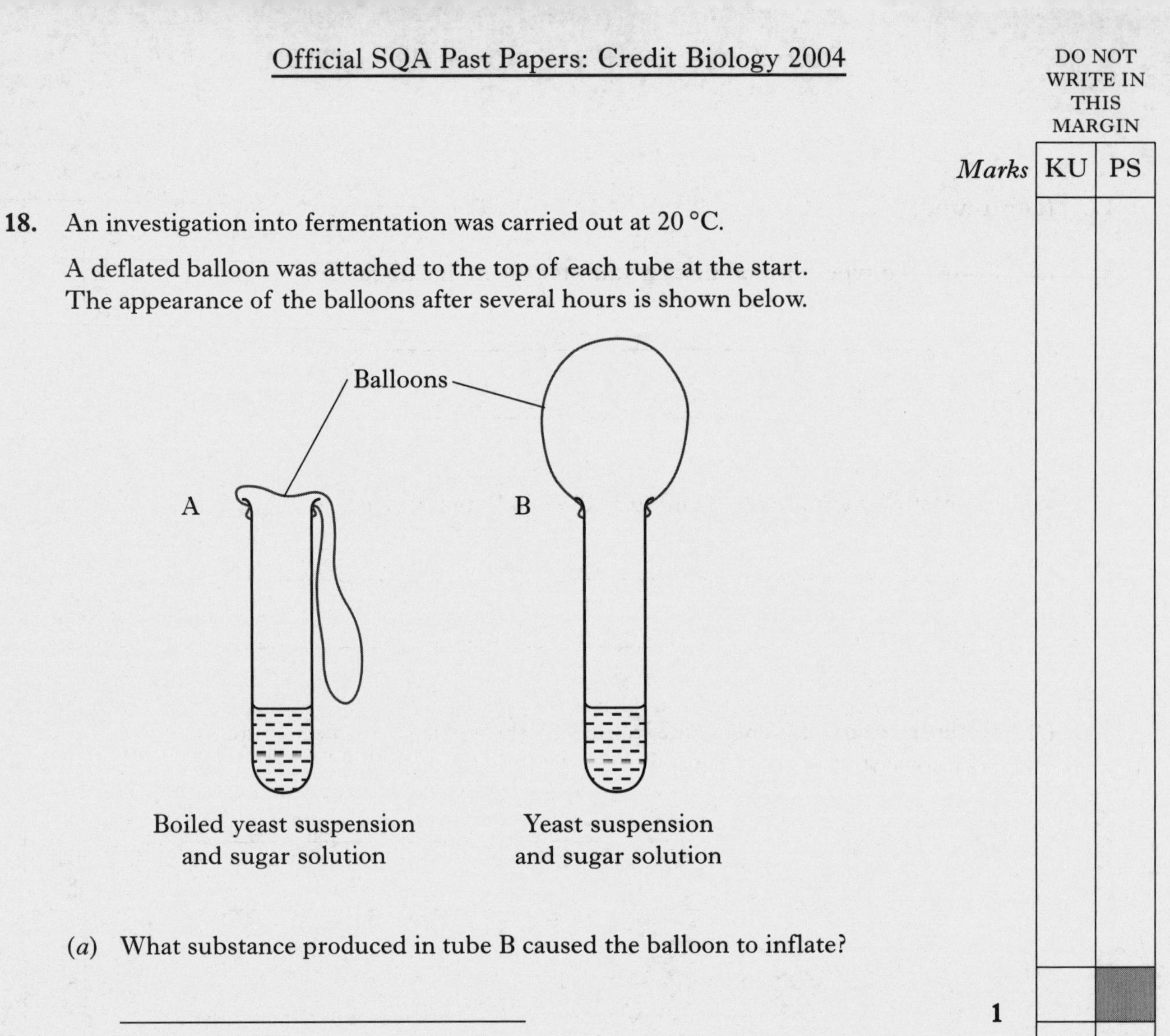

(*a*) What substance produced in tube B caused the balloon to inflate?

______________________________ **1**

(*b*) The balloon on the tube with the boiled yeast suspension did not inflate. Explain this result.

____

______________________________ **1**

(*c*) How would the appearance of the balloon on tube B differ if the investigation had been carried out at 15 °C?

______________________________ **1**

DO NOT WRITE IN THIS MARGIN

Marks | KU | PS

19. The diagrams show the production of insulin by genetic engineering.
They are not in the correct order.

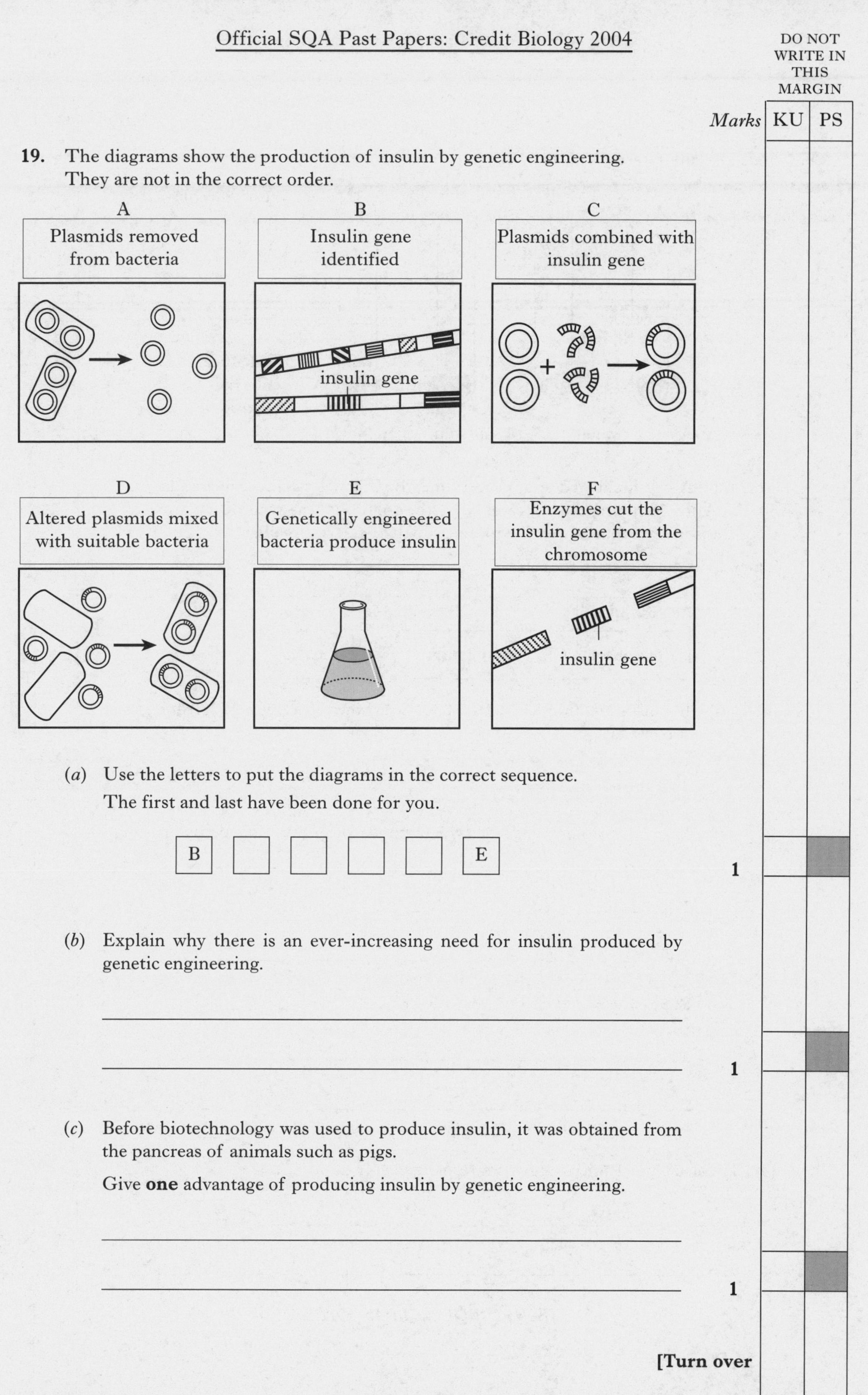

(*a*) Use the letters to put the diagrams in the correct sequence.
The first and last have been done for you.

B ☐ ☐ ☐ ☐ E **1**

(*b*) Explain why there is an ever-increasing need for insulin produced by genetic engineering.

____

____ **1**

(*c*) Before biotechnology was used to produce insulin, it was obtained from the pancreas of animals such as pigs.

Give **one** advantage of producing insulin by genetic engineering.

____

____ **1**

[Turn over

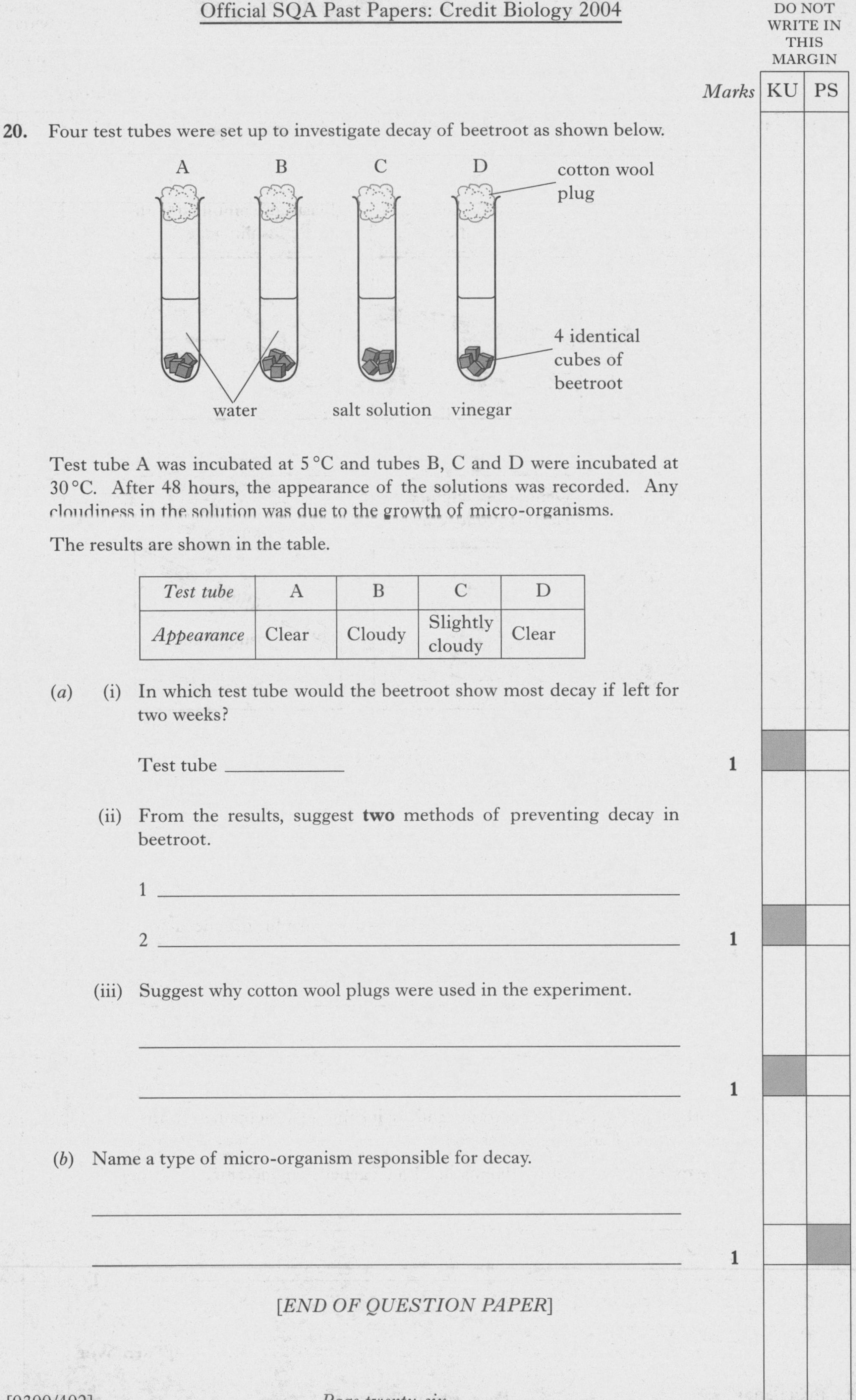

Marks | KU | PS

20. Four test tubes were set up to investigate decay of beetroot as shown below.

Test tube A was incubated at 5 °C and tubes B, C and D were incubated at 30 °C. After 48 hours, the appearance of the solutions was recorded. Any cloudiness in the solution was due to the growth of micro-organisms.

The results are shown in the table.

Test tube	A	B	C	D
Appearance	Clear	Cloudy	Slightly cloudy	Clear

(*a*) (i) In which test tube would the beetroot show most decay if left for two weeks?

Test tube __________ **1**

(ii) From the results, suggest **two** methods of preventing decay in beetroot.

1 ____________________

2 ____________________ **1**

(iii) Suggest why cotton wool plugs were used in the experiment.

____

____________________ **1**

(*b*) Name a type of micro-organism responsible for decay.

____

____________________ **1**

[*END OF QUESTION PAPER*]

ADDITIONAL GRAPH PAPER FOR QUESTION 16(*b*)(i)

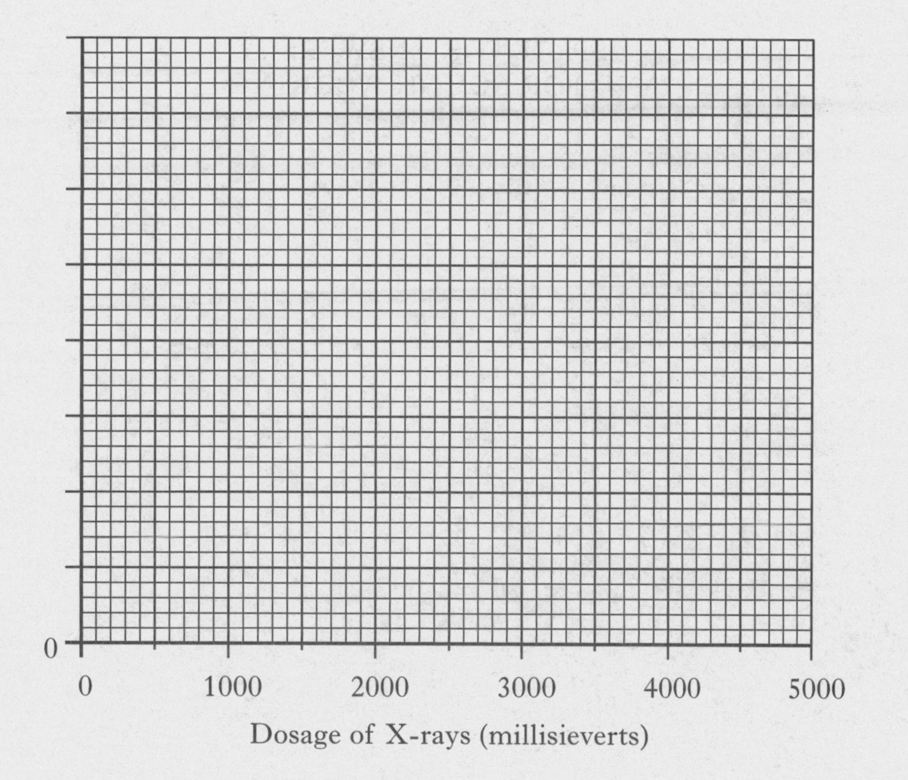

[Turn over

SPACE FOR ANSWERS
AND FOR ROUGH WORKING

2005 | Credit

[BLANK PAGE]

FOR OFFICIAL USE

C

KU	PS

Total Marks

0300/402

NATIONAL QUALIFICATIONS 2005

WEDNESDAY, 18 MAY
10.50 AM – 12.20 PM

BIOLOGY
STANDARD GRADE
Credit Level

Fill in these boxes and read what is printed below.

Full name of centre

Town

Forename(s)

Surname

Date of birth
Day Month Year

Scottish candidate number

Number of seat

1 All questions should be attempted.

2 The questions may be answered in any order but all answers are to be written in the spaces provided in this answer book, and must be written clearly and legibly in ink.

3 Rough work, if any should be necessary, as well as the fair copy, is to be written in this book. Additional spaces for answers and for rough work will be found at the end of the book. Rough work should be scored through when the fair copy has been written.

4 Before leaving the examination room you must give this book to the invigilator. If you do not, you may lose all the marks for this paper.

LIB 0300/402 6/29670

DO NOT WRITE IN THIS MARGIN

	Marks	KU	PS

1. (*a*) Use information from the diagrams of invertebrates to complete the following paired statement key.

The diagrams are not to the same scale.

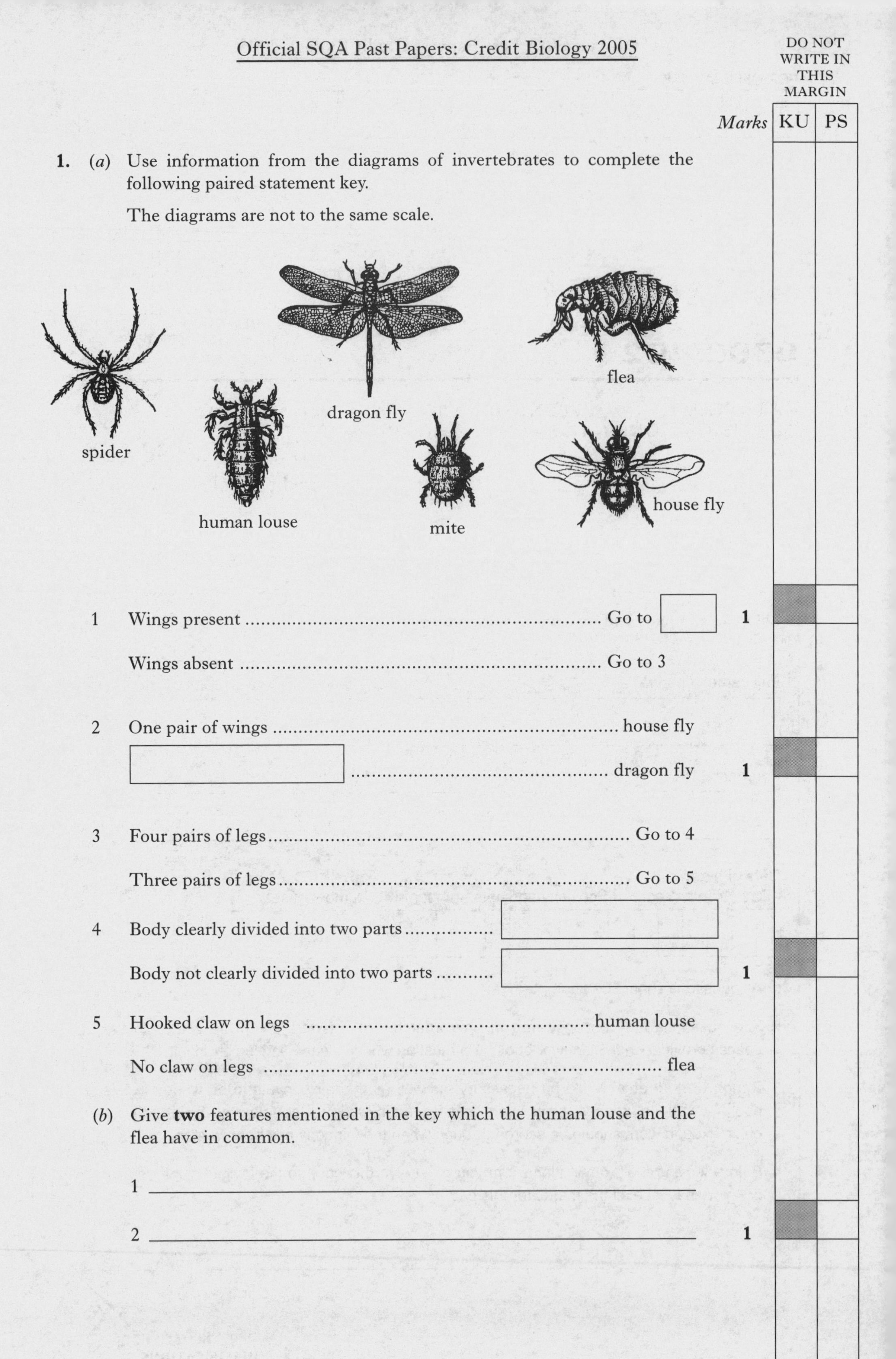

1 Wings present .... Go to [ ] **1**

Wings absent .... Go to 3

2 One pair of wings .... house fly

[ ] .... dragon fly **1**

3 Four pairs of legs .... Go to 4

Three pairs of legs .... Go to 5

4 Body clearly divided into two parts .... [ ]

Body not clearly divided into two parts .... [ ] **1**

5 Hooked claw on legs .... human louse

No claw on legs .... flea

(*b*) Give **two** features mentioned in the key which the human louse and the flea have in common.

1 ____

2 ____ **1**

Marks KU PS

2. The light intensity inside and outside a woodland was measured over a year. The table shows the results.

	Average daily light intensity (units)	
Month	*Outside woodland*	*Inside woodland*
January	10	8
February	13	10
March	15	12
April	19	16
May	24	22
June	28	15
July	30	5
August	25	5
September	20	5
October	15	5
November	12	10
December	10	8

(*a*) On the grid below, complete the Y axis, key and line graph plot to show the results.
(An additional grid, if required, will be found on page 22.)

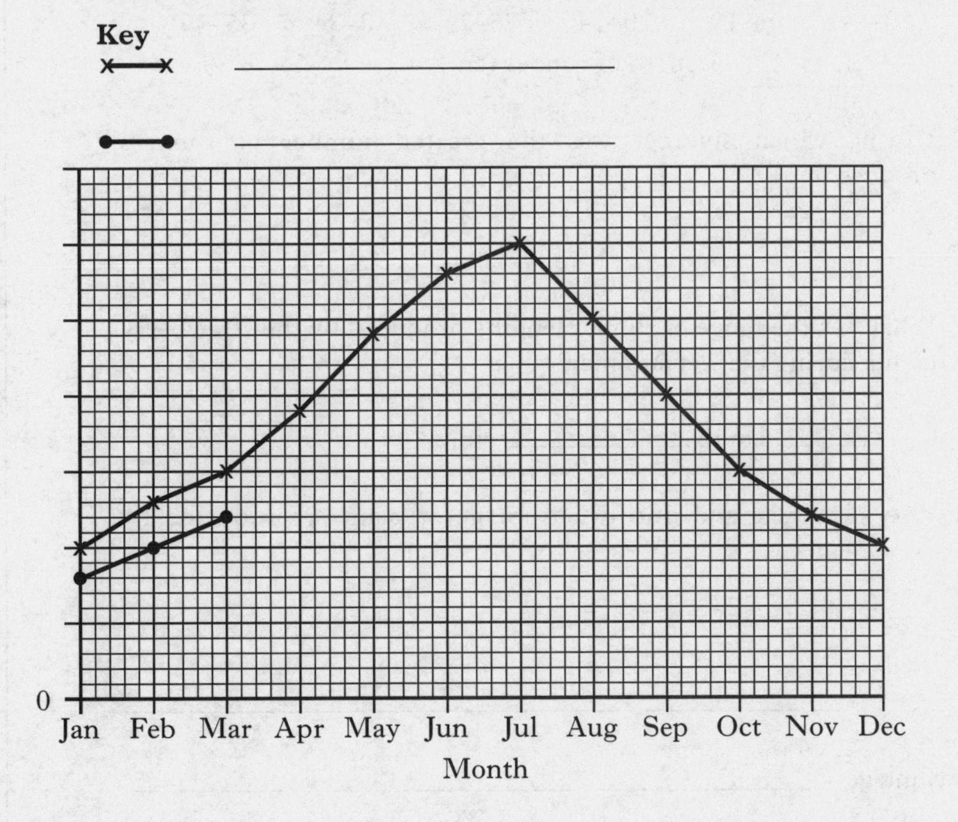

2

(*b*) Explain why the difference between the light intensities inside and outside the woodland is much greater from June to October.

1

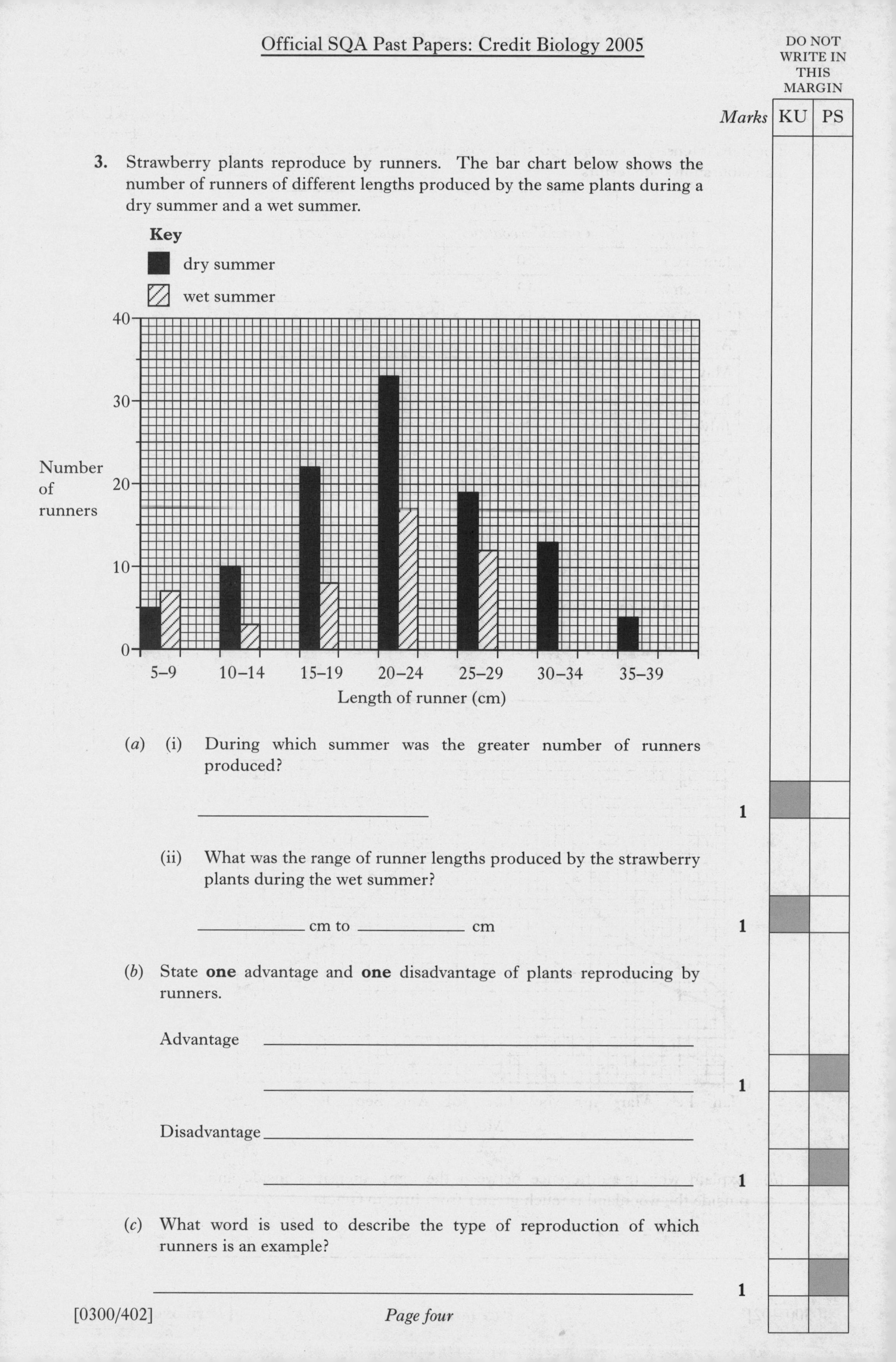

DO NOT WRITE IN THIS MARGIN

Marks KU PS

3. Strawberry plants reproduce by runners. The bar chart below shows the number of runners of different lengths produced by the same plants during a dry summer and a wet summer.

(*a*) (i) During which summer was the greater number of runners produced?

________________________ **1**

(ii) What was the range of runner lengths produced by the strawberry plants during the wet summer?

__________ cm to __________ cm **1**

(*b*) State **one** advantage and **one** disadvantage of plants reproducing by runners.

Advantage ____________________________________

____________________________________ **1**

Disadvantage ____________________________________

____________________________________ **1**

(*c*) What word is used to describe the type of reproduction of which runners is an example?

____________________________________ **1**

DO NOT WRITE IN THIS MARGIN

Marks | KU | PS

3. (continued)

(*d*) Plants may also produce seeds which can be dispersed away from the parent plant. The diagrams below show some seeds and fruits of named plants. They all use one of two methods of seed dispersal.

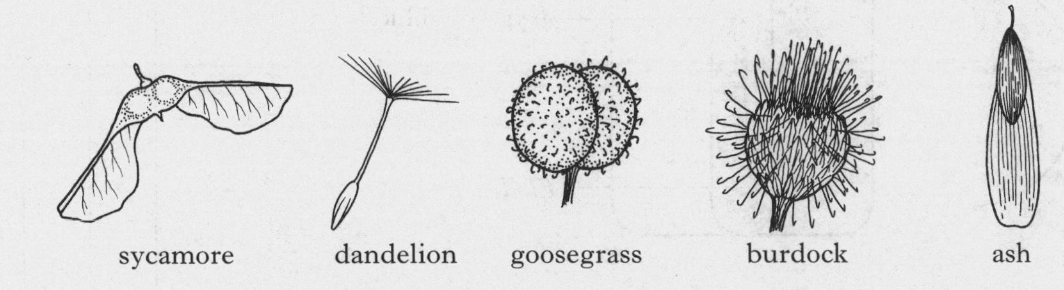

(i) Complete the following table to identify each of these methods of seed dispersal and the plants which use them.

Method of seed dispersal	*Plants which use this method*

2

(ii) Plants with succulent fruits use a different method of seed dispersal.

Describe this method.

____

____ 1

[Turn over

DO NOT WRITE IN THIS MARGIN

Marks | KU | PS

4. In an investigation into the rate of photosynthesis, a piece of *Elodea* (pondweed) was placed in a beaker of water and a bright light shone on it.

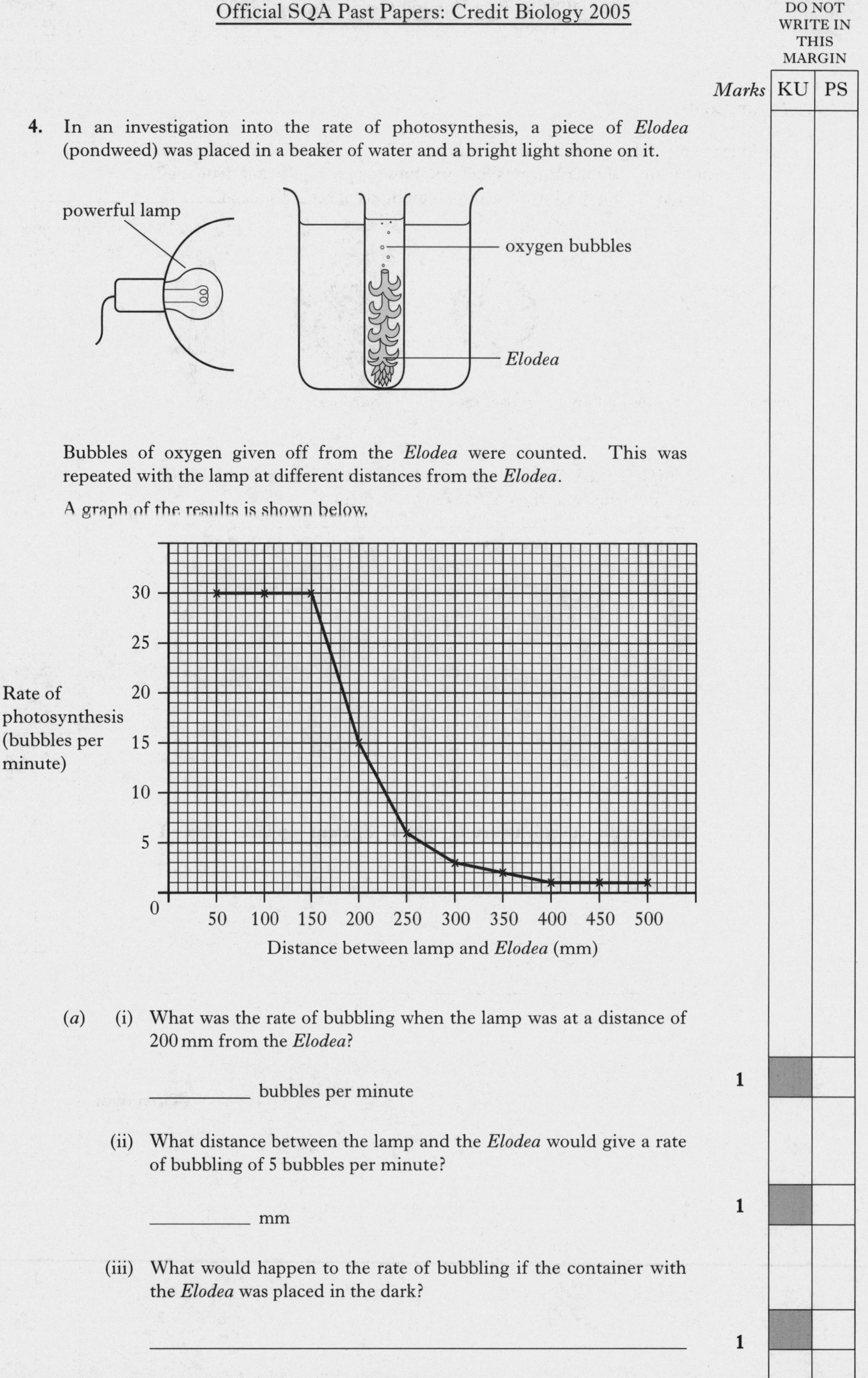

Bubbles of oxygen given off from the *Elodea* were counted. This was repeated with the lamp at different distances from the *Elodea*.

A graph of the results is shown below.

(*a*) (i) What was the rate of bubbling when the lamp was at a distance of 200 mm from the *Elodea*?

____________ bubbles per minute **1**

(ii) What distance between the lamp and the *Elodea* would give a rate of bubbling of 5 bubbles per minute?

____________ mm **1**

(iii) What would happen to the rate of bubbling if the container with the *Elodea* was placed in the dark?

____ **1**

DO NOT WRITE IN THIS MARGIN

Marks | KU | PS

4. (continued)

(*b*) (i) At which distances between the lamp and the *Elodea* did light act as a limiting factor on the rate of photosynthesis?
Tick the correct box.

50 – 150 mm ☐

150 – 400 mm ☐

400 – 500 mm ☐

1

(ii) Name **one** other factor which could limit the rate of photosynthesis.

____ **1**

(*c*) The investigation was carried out several times and the average results were used to plot the graph. Why was this good experimental technique?

____ **1**

[Turn over

DO NOT WRITE IN THIS MARGIN

Marks | KU | PS

5. The water quality at beaches is tested to check that it is not affected by any untreated sewage.

The table gives information about the number of beaches which were tested in one particular year and the number passed as suitable for swimming.

Country	*Number of beaches tested*	*Number of beaches suitable for swimming*	*Percentage of beaches suitable for swimming*
England	271	239	88·2
Scotland	93	68	
Wales	128	102	79·7
Northern Ireland	17	16	94·1

(*a*) Complete the table to show the percentage of beaches suitable for swimming in Scotland.

Space for calculation

1

(*b*) Why should the percentages of beaches which passed be used when comparing the results from the four countries, rather than the actual number?

1

(*c*) The samples of water from the beaches can be examined for the presence of certain species. This technique gives information about water pollution. What name is given to such species?

1

DO NOT WRITE IN THIS MARGIN

Marks | KU | PS

6. The diagram below shows part of the nitrogen cycle.

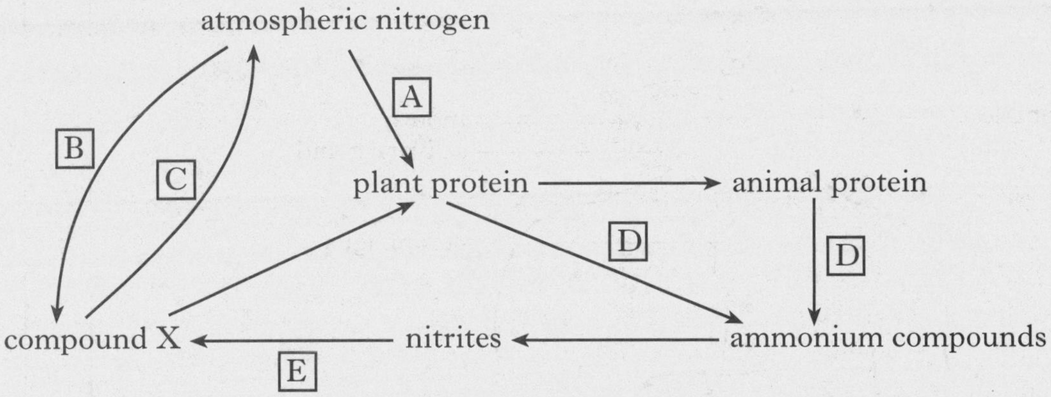

atmospheric nitrogen
A
B
C
plant protein
animal protein
D
D
compound X
nitrites
ammonium compounds
E

(a) Use letters from the diagram to complete the following table about some of the events of the nitrogen cycle.

Event	*Letter*
Death and decay	
Action by denitrifying bacteria	
Lightning	

2

(b) Explain why event A can take place in some plants such as clover, peas and beans, but not in others.

____

____ **1**

(c) Name compound X.

____________________ **1**

[Turn over

DO NOT WRITE IN THIS MARGIN

Marks | KU | PS

7. The following diagram shows the human digestive system.

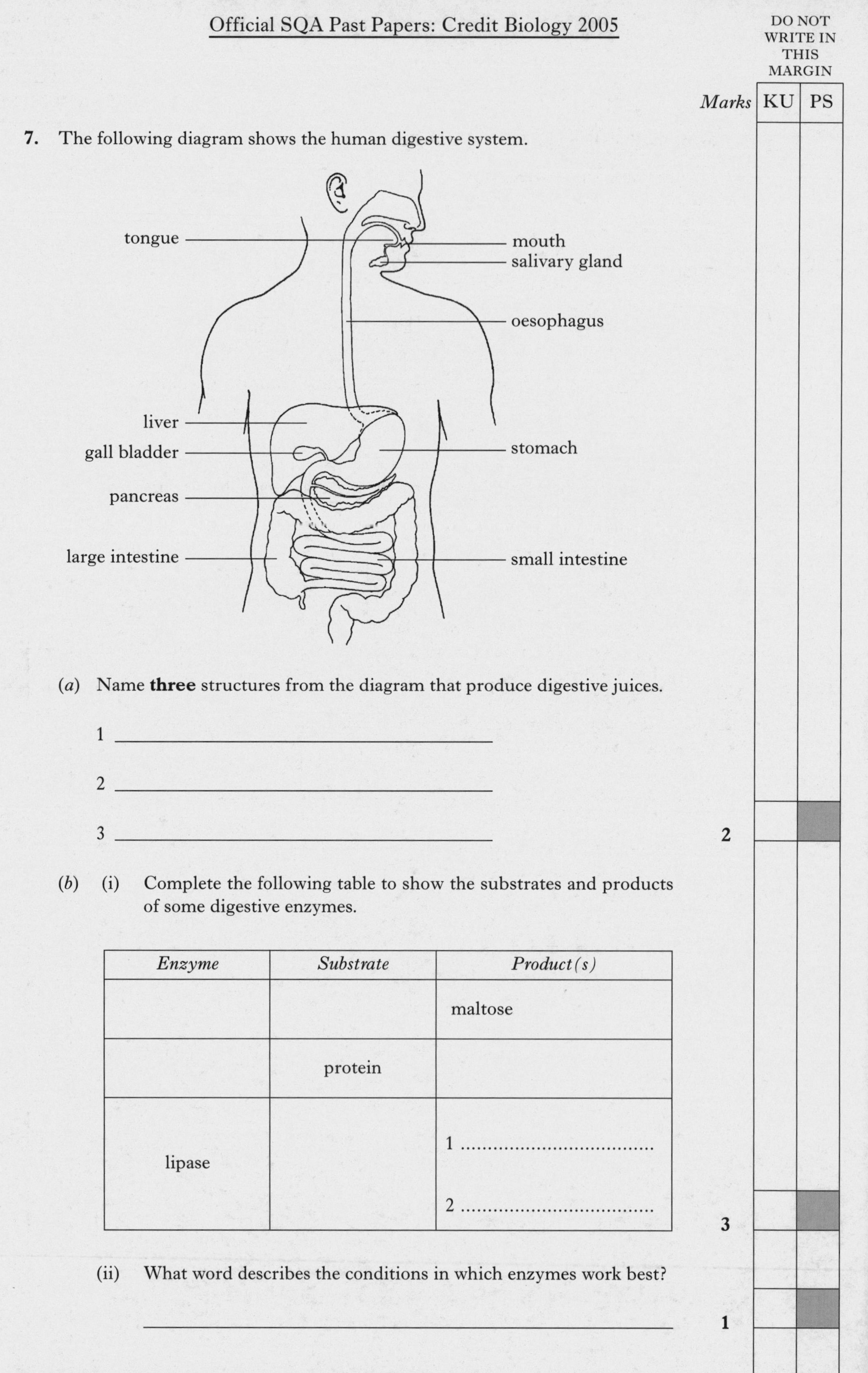

(*a*) Name **three** structures from the diagram that produce digestive juices.

1 ______________________

2 ______________________

3 ______________________ **2**

(*b*) (i) Complete the following table to show the substrates and products of some digestive enzymes.

Enzyme	*Substrate*	*Product(s)*
		maltose
	protein	
lipase		1 .... 2 ....

3

(ii) What word describes the conditions in which enzymes work best?

______________________ **1**

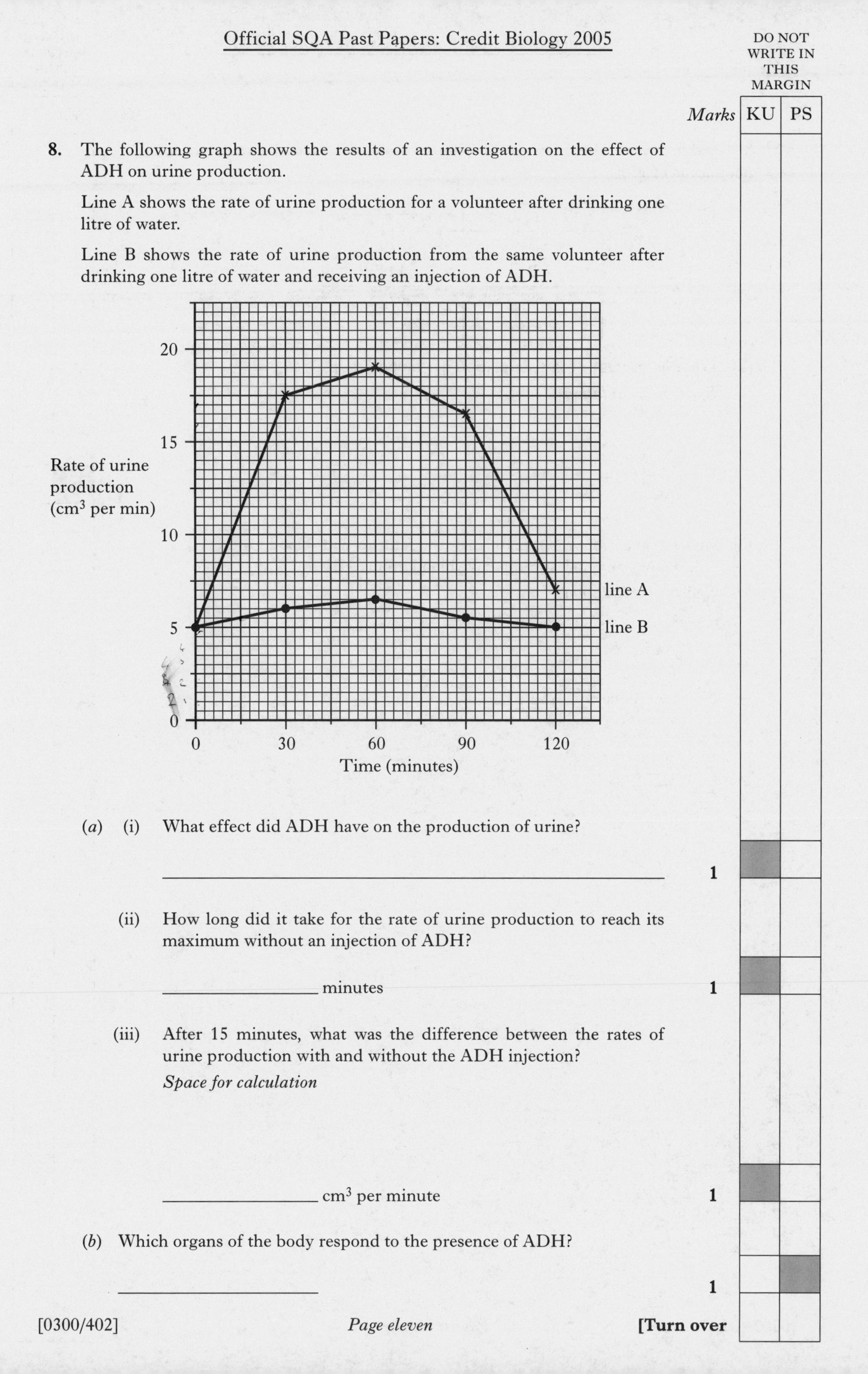

DO NOT WRITE IN THIS MARGIN

Marks	KU	PS

8. The following graph shows the results of an investigation on the effect of ADH on urine production.

Line A shows the rate of urine production for a volunteer after drinking one litre of water.

Line B shows the rate of urine production from the same volunteer after drinking one litre of water and receiving an injection of ADH.

(*a*) (i) What effect did ADH have on the production of urine?

______________________________ **1**

(ii) How long did it take for the rate of urine production to reach its maximum without an injection of ADH?

______________ minutes **1**

(iii) After 15 minutes, what was the difference between the rates of urine production with and without the ADH injection?

Space for calculation

______________ cm^3 per minute **1**

(*b*) Which organs of the body respond to the presence of ADH?

______________ **1**

DO NOT WRITE IN THIS MARGIN

Marks KU PS

9. The table shows the number of people with each blood group in a population of 1500.

Blood group	*Number of people*
A	610
B	143
O	675
AB	72

(*a*) What percentage of the population has blood group O?

Space for calculation

_____ % **1**

(*b*) In the population, the ratio of males to females with blood group AB is 5:3. How many males would have blood group AB?

Space for calculation

Number of males _________ **1**

DO NOT WRITE IN THIS MARGIN

Marks KU PS

9. (continued)

(*c*) Blood platelets are important in the formation of blood clots at the site of an injury. The following diagram shows the sequence of reactions which produce the clot when platelets gather at the injury.

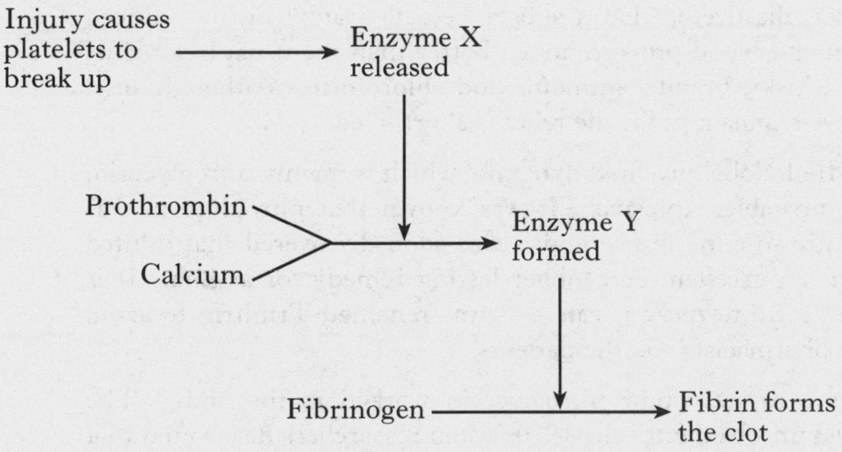

Injury causes platelets to break up → Enzyme X released

Prothrombin + Calcium (+ Enzyme X) → Enzyme Y formed

Fibrinogen (+ Enzyme Y) → Fibrin forms the clot

(i) Suggest **one** benefit of a blood clot forming at the site of an injury.

____ **1**

(ii) Explain why low calcium levels would reduce the blood's ability to clot.

____

____ **1**

[Turn over

DO NOT WRITE IN THIS MARGIN

Marks | KU | PS

10. Read the following passage and answer the questions based on it.

An Explosive Medication

Angina is a pain in the chest that happens when the heart muscle does not receive enough blood. It occurs when branches of the artery that carries blood to the heart become narrowed or blocked. In 1867, T. L. Brunton, an Edinburgh doctor, discovered that a substance called amyl nitrite reduced both angina pain and blood pressure much better than the usual treatments which included whisky, brandy, ammonia and chloroform. Although amyl nitrite does relieve angina rapidly, the relief is short lived.

Also in 1867, Alfred Nobel invented dynamite which is mainly nitroglycerin, a powerful and unstable explosive. It was known that nitroglycerin was similar in structure to amyl nitrite and it was soon discovered that diluted nitroglycerin was an excellent and longer lasting remedy for angina. It is used in a diluted form to make it safe and was renamed Trinitrin to avoid scaring both the pharmacists and the patients.

However, it was a mystery how nitroglycerin worked in the body. The mystery remained unsolved until the 1970s when researchers discovered that it works by changing into nitric oxide. Outside the body, nitric oxide is a poisonous gas but it plays a vital part inside the body. Nitric oxide is the main messenger making blood vessels open wider so that more blood flows to the starved heart muscle, and this is why nitroglycerin helps angina patients.

(*a*) What is the name of the artery which carries blood to the heart muscle?

____________________________ **1**

(*b*) What is the cause of the chest pain referred to as angina?

____ **1**

(*c*) What made nitroglycerin appear suitable for investigating as a possible heart medicine?

____ **1**

(*d*) Nitroglycerin is diluted to make it safe. Why is this necessary?

____ **1**

(*e*) Complete the following flow diagram to show how nitroglycerin works in the body.

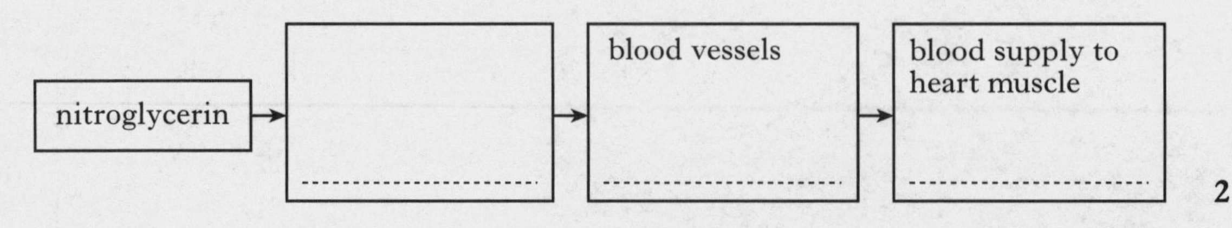

2

DO NOT WRITE IN THIS MARGIN

Marks | KU | PS

11. The diagrams below show some of the muscles in the leg.

position 1 — X, Y

position 2 — X, Y

(*a*) (i) Describe the action of muscles X and Y as the leg moves from position 1 to position 2.

Muscle X ______________________________

Muscle Y ______________________________ **1**

(ii) Name the structures which attach muscles to bones.

______________________________ **1**

(*b*) (i) Draw a line from each of the following parts of the brain to its correct function.

Part	*Function*
cerebrum	controls heart rate
cerebellum	controls balance
medulla	enables conscious thought and memory

2

(ii) Underline **one** option in each bracket to describe the flow of information in the nervous system.

Information from the environment is detected by the { heart / sense organs / brain } and sent to the { central nervous system / circulation system / skin } which responds by sending messages to the { muscles / blood / bones }. **1**

DO NOT WRITE IN THIS MARGIN

Marks | KU | PS

12. The table gives the average yield of maize (corn) per hectare in the USA since 1960.

Year	1960	1970	1980	1990	2000
Average annual yield (tonnes per hectare)	4·05	5·32	7·14	8·21	10·05

(*a*) (i) Calculate the average annual increase in yield in the 40 year period from 1960 to 2000.

Space for calculation

Average annual increase in yield __________ tonnes per hectare **1**

(ii) Which ten year period showed the greatest average increase in yield?

Tick the correct box.

Space for calculation

1960 – 1970 ☐ 1980 – 1990 ☐

1970 – 1980 ☐ 1990 – 2000 ☐ **1**

(*b*) Each individual plant in the field gives a different yield which can be any value between the lowest and the highest.

What name is given to this type of variation?

________________________________ **1**

DO NOT WRITE IN THIS MARGIN

Marks	KU	PS

12. (continued)

(*c*) (i) The improvement in yield has been largely due to the production of new varieties of maize by *selective breeding*.

Explain what is meant by this term.

____

____

____ **1**

(ii) It is possible to produce new varieties of maize by introducing genes from species which do not interbreed with maize.

What general name is given to these techniques?

____________________________ **1**

(*d*) In order to increase the variation available for selective breeding, plant biologists treat maize in ways that can increase the rate of mutation.

(i) What is meant by the term *mutation*?

____ **1**

(ii) Give an example of a factor that can increase the rate of mutation.

____________________________ **1**

(*e*) Farmers try to ensure the maximum yield of crops. This requires a plentiful supply of plant nutrients and little competition from other plants.

Describe how each of these can be achieved.

Plentiful supply of plant nutrients ________________________

____ **1**

Reduced competition from other plants ________________________

____ **1**

[Turn over

DO NOT WRITE IN THIS MARGIN

Marks | KU | PS

13. The table shows a comparison of the breakdown of one gram of glucose by three different types of cell respiration.

	Type of cell respiration		
	A	**B**	**C**
Energy released (kJ)	17·1	0·9	0·9
Oxygen used (g)	1·07	0	0
Carbon dioxide produced (g)	1·47	0	0·49
Water produced (g)	0·6	0	0
Lactic acid produced (g)	0	1	0
Ethanol produced (g)	0	0	0·51

(*a*) (i) Respiration of type A releases much more energy than the other types.

What name is given to this type of respiration?

______________________________ **1**

(ii) Which **two** types of respiration take place in the following cells?

Muscle cells: type ______ and type ______ **1**

Yeast cells: type ______ and type ______ **1**

(*b*) Express the energy released from one gram of glucose by the three types of respiration as a simple whole number ratio.

Space for calculation

______ : ______ : ______ **1**

type A type B type C

(*c*) Give **one** way in which the chemical energy released from food is important in the metabolism of cells.

____

____ **1**

DO NOT WRITE IN THIS MARGIN

Marks | KU | PS

14. In an investigation on gas exchange, samples of breathed air were collected from several volunteers. The table shows the volumes of carbon dioxide and oxygen in 1000 cm^3 of each sample.

Sample	*Volume of carbon dioxide* (cm^3)	*Volume of oxygen* (cm^3)
A	10	153
B	7	148
C	6	154
D	11	153
E	6	152
Average	8	

(*a*) Complete the table by calculating the average volume of oxygen in the samples.

Space for calculation

1

(*b*) Calculate the percentage of oxygen in sample C.

Space for calculation

_______ %

1

(*c*) Name the chemical in the blood which combines with oxygen to transport it to the body tissues.

____

1

[Turn over

DO NOT WRITE IN THIS MARGIN

Marks | KU | PS

15. The table shows the composition of 100 g of four common fruits.

	Component			
Fruit	*Protein* (g)	*Carbohydrate* (g)	*Fat* (g)	*Water* (g)
bananas	1·0	23	0·3	
apples	0·4	12	0·1	87·5
pears	0·4	10	0·1	89·5
grapes	0·2	15	0·1	

(*a*) Complete the table by adding the mass of water present in 100 g of bananas and grapes.

Space for calculation

1

(*b*) Which component of the fruits contains the most energy per gram?

____________________ **1**

(*c*) Give **two** differences between the composition of apples and that of pears.

1 ____________________

2 ____________________ **2**

(*d*) Suggest how it would be possible to minimise the effect of variations of individual fruits when measuring their composition.

____

____________________ **1**

DO NOT WRITE IN THIS MARGIN

Marks | KU | PS

16. Identical pieces of cloth were marked with stains. They were then washed at different temperatures using biological detergent. The degree of staining still on the cloth after washing was measured and expressed as a percentage of the stain before washing. The test was repeated using a non-biological detergent.

The results are shown in the graph below.

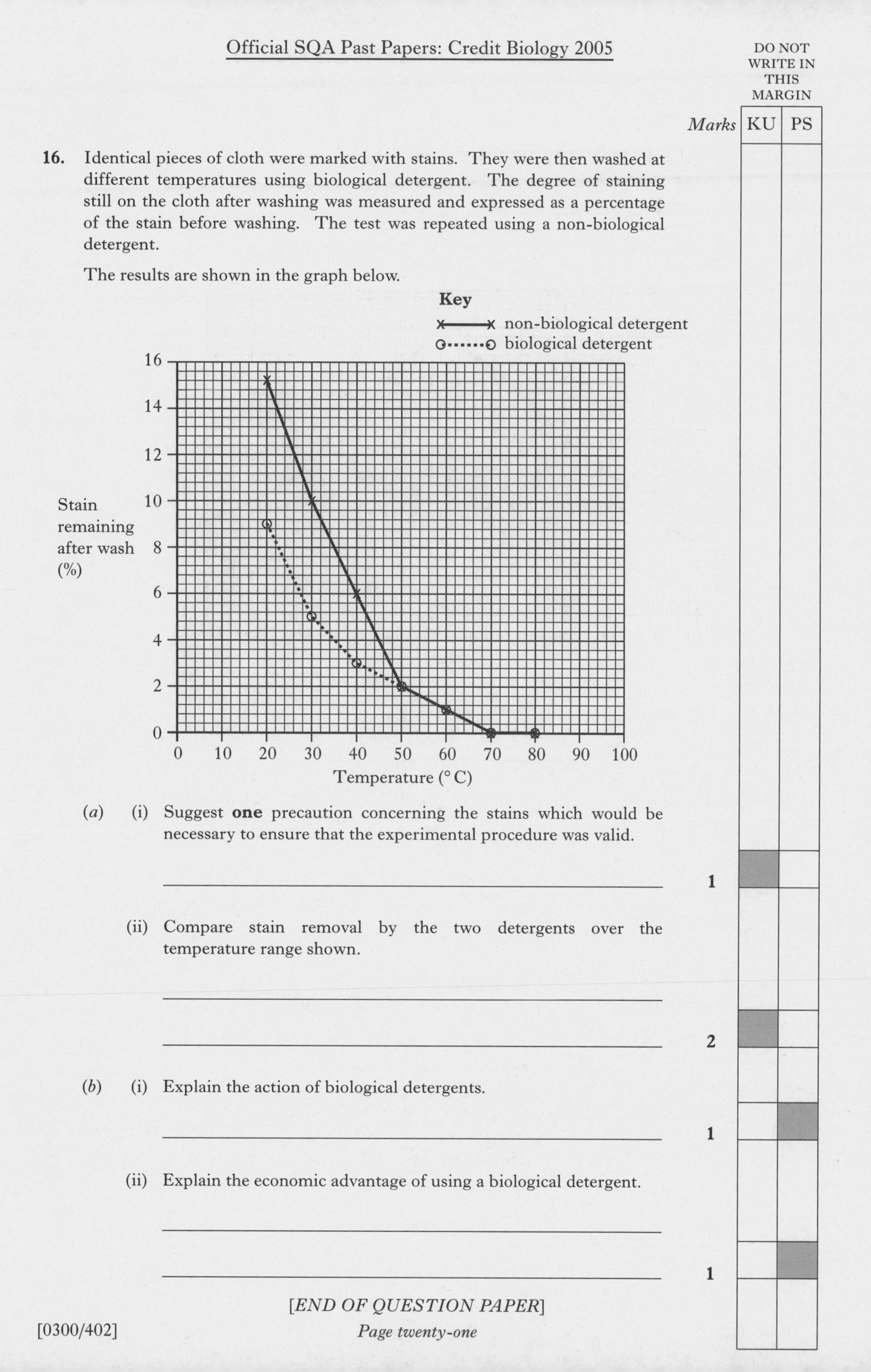

(*a*) (i) Suggest **one** precaution concerning the stains which would be necessary to ensure that the experimental procedure was valid.

____ **1**

(ii) Compare stain removal by the two detergents over the temperature range shown.

____

____ **2**

(*b*) (i) Explain the action of biological detergents.

____ **1**

(ii) Explain the economic advantage of using a biological detergent.

____

____ **1**

[*END OF QUESTION PAPER*]

ADDITIONAL GRAPH PAPER FOR QUESTION 2(*a*)

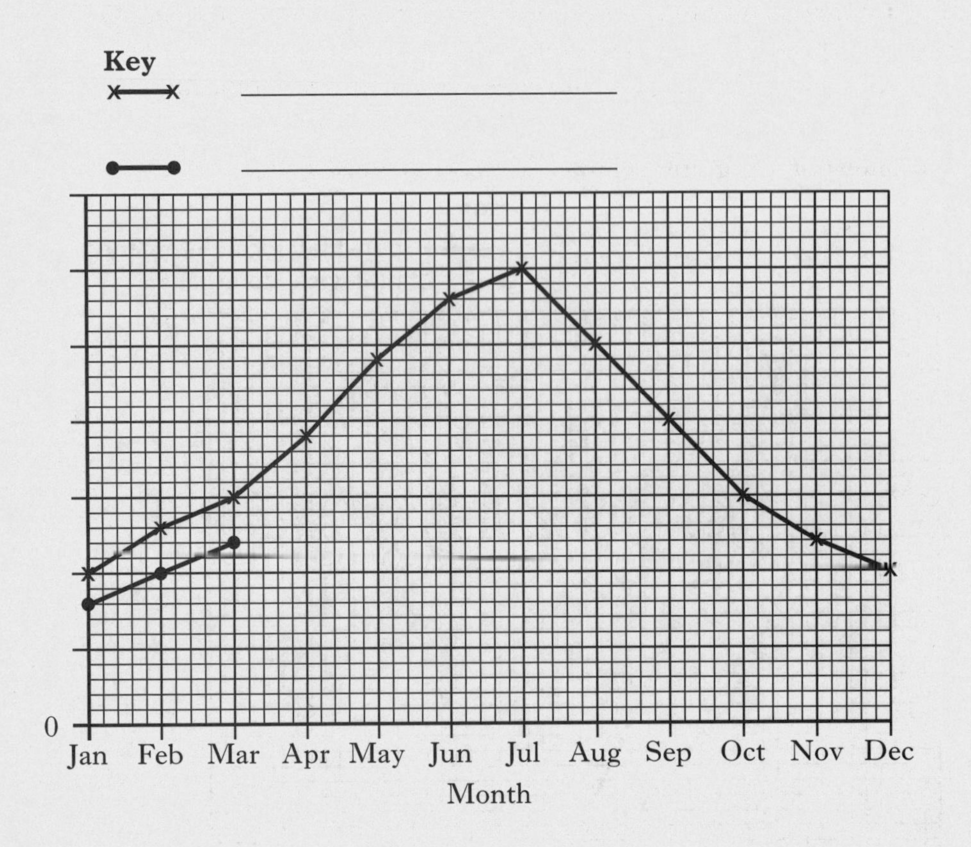

2006 | Credit

[BLANK PAGE]

FOR OFFICIAL USE

C

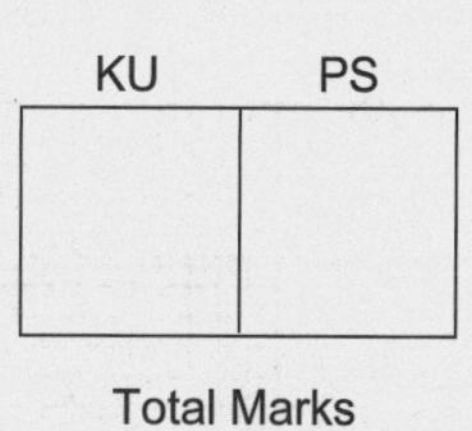

0300/402

NATIONAL
QUALIFICATIONS
2006

TUESDAY, 23 MAY
10.50 AM – 12.20 PM

BIOLOGY
STANDARD GRADE
Credit Level

Fill in these boxes and read what is printed below.

Full name of centre

Town

Forename(s)

Surname

Date of birth
Day Month Year

Scottish candidate number

Number of seat

1 All questions should be attempted.

2 The questions may be answered in any order but all answers are to be written in the spaces provided in this answer book, and must be written clearly and legibly in ink.

3 Rough work, if any should be necessary, as well as the fair copy, is to be written in this book. Additional spaces for answers and for rough work will be found at the end of the book. Rough work should be scored through when the fair copy has been written.

4 Before leaving the examination room you must give this book to the invigilator. If you do not, you may lose all the marks for this paper.

SCOTTISH
QUALIFICATIONS
AUTHORITY

DO NOT WRITE IN THIS MARGIN

Marks KU PS

1. The graph shows the growth curve of a population of bacteria in a fermenter at 30 °C over a 24 hour period.

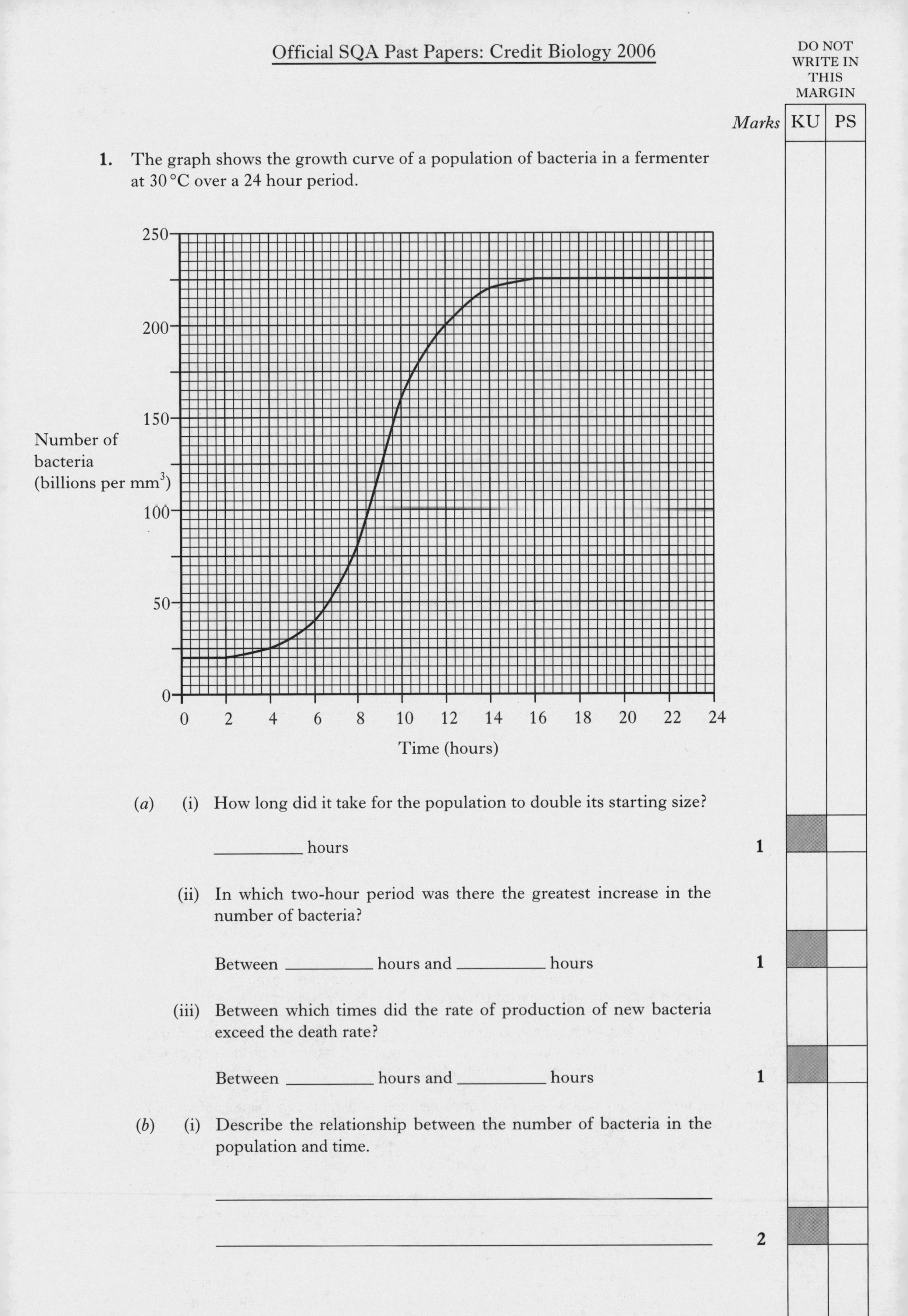

(*a*) (i) How long did it take for the population to double its starting size?

__________ hours **1**

(ii) In which two-hour period was there the greatest increase in the number of bacteria?

Between __________ hours and __________ hours **1**

(iii) Between which times did the rate of production of new bacteria exceed the death rate?

Between __________ hours and __________ hours **1**

(*b*) (i) Describe the relationship between the number of bacteria in the population and time.

____

____ **2**

DO NOT WRITE IN THIS MARGIN

Marks KU PS

1. (*b*) (continued)

(ii) Explain the changes in the shape of the growth curve between 10 hours and 18 hours in terms of the factors that affect population growth.

____

____ **2**

(iii) Draw a second line on the graph to predict the growth in population if the fermenter had been kept at a temperature of 20 °C.

(An additional graph, if needed, will be found on page 22.) **1**

(*c*) (i) Before setting up the fermenter, all the apparatus was heated to 150 °C for 15 minutes to eliminate any contamination by resistant cells of bacteria and fungi.

What name is given to these resistant cells?

________________________ **1**

(ii) The fermenter was stirred before removing the samples used to estimate the numbers of bacteria. How would this minimise possible errors in the results?

____

____ **1**

[Turn over

DO NOT WRITE IN THIS MARGIN

Marks | KU | PS

2. (*a*) The table shows the percentage germination of four crop plants over a range of temperatures.

Temperature (°C)	*Percentage germination of crop plants*			
	Carrots	*Cauliflower*	*Okra*	*Spinach*
0	0	0	0	83
5	48	0	0	96
10	93	58	0	91
15	95	60	74	80
20	96	65	89	52
25	95	53	93	28
30	90	45	88	14
35	74	0	85	0
40	0	0	35	0

(i) Which **two** crop plants are able to germinate over the widest range of temperatures?

1 ______________________ 2 ______________________ **1**

(ii) Complete the table below by adding the correct heading, units and values to show the optimum germination temperature for each of the crop plants.

Crop plant	

2

(iii) Suggest which crop plant would germinate best in a hot climate.

______________________ **1**

DO NOT WRITE IN THIS MARGIN

Marks | KU | PS

2. (*a*) (continued)

(iv) What is the minimum number of spinach seeds which should be sown at 15 °C in order to produce 1000 seedling plants?

Space for calculation

Number of seeds ___________ **1**

(v) On the grid below, complete a line graph of the change in percentage germination of cauliflower seeds with temperature.

(An additional graph, if needed, will be found on page 22.)

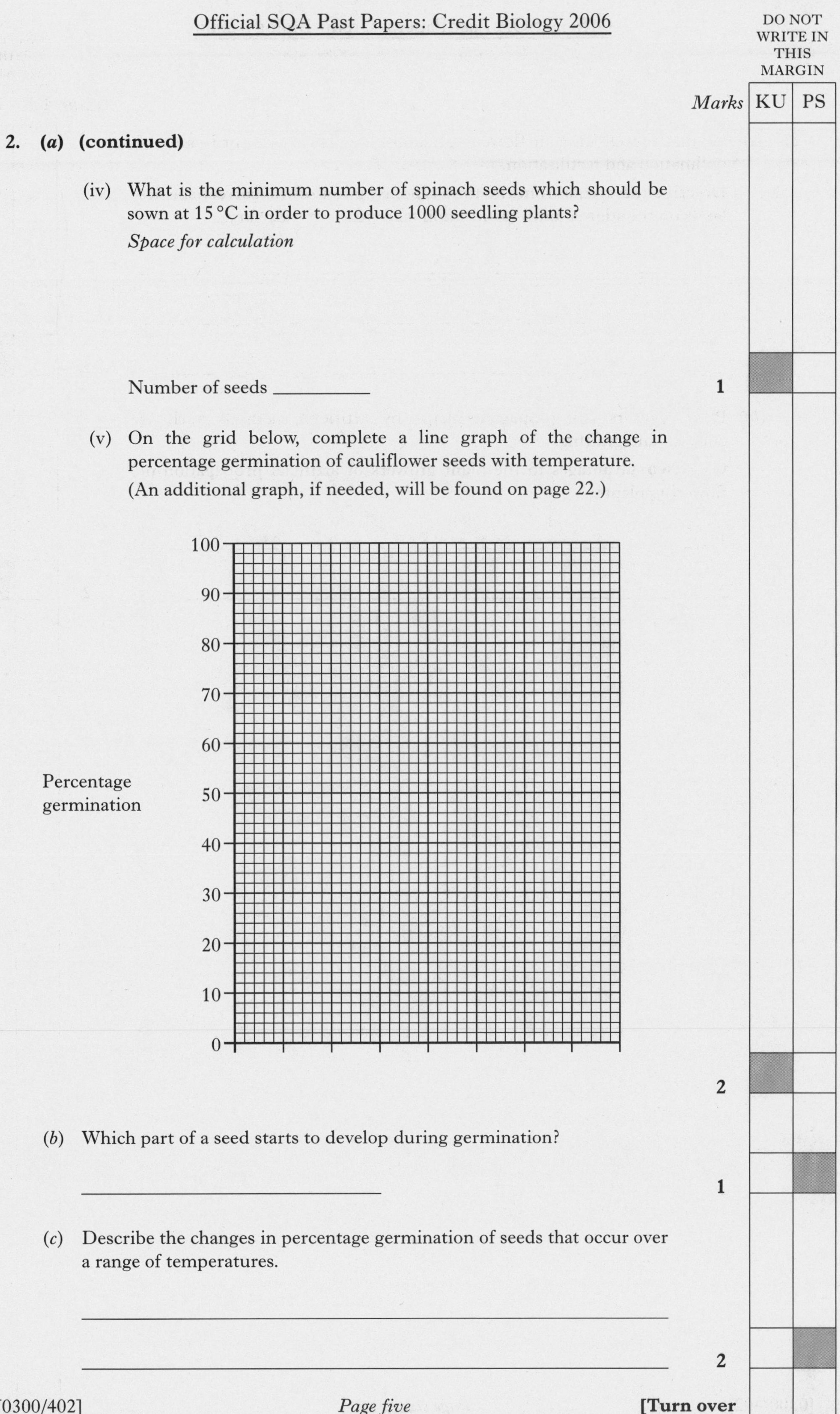

2

(*b*) Which part of a seed starts to develop during germination?

______________________________ **1**

(*c*) Describe the changes in percentage germination of seeds that occur over a range of temperatures.

____

______________________________ **2**

DO NOT WRITE IN THIS MARGIN

Marks | KU | PS

3. (*a*) Sexual reproduction in flowering plants depends on the processes of pollination and fertilisation.

Describe the events from the time a pollen grain of the correct species lands on the stigma, until fertilisation takes place in the ovary.

____

____

______________________________ **2**

(*b*) Plant growers can propagate plants by artificial methods such as cuttings and grafting.

Give **two** advantages to the plant growers of artificial propagation of flowering plants.

1 ______________________________

2 ______________________________ **2**

DO NOT WRITE IN THIS MARGIN

Marks KU PS

4. (*a*) Fertilisation is the fusion of gametes and can be either internal or external in animals.

Explain why it is necessary for some animals to use internal fertilisation.

____

____ **1**

(*b*) A human fetus develops inside the mother's uterus, attached to the placenta.

Name **one** substance which passes across the placenta from mother to fetus.

____________________________________ **1**

(*c*) Some animal species take more care of their young than others.

Describe the relationship between the degree of parental care and the number of eggs that are produced at any one time by different species.

____

____ **1**

[Turn over

DO NOT WRITE IN THIS MARGIN

Marks KU PS

5. (*a*) Food is moved along the alimentary canal by the action of circular muscles.

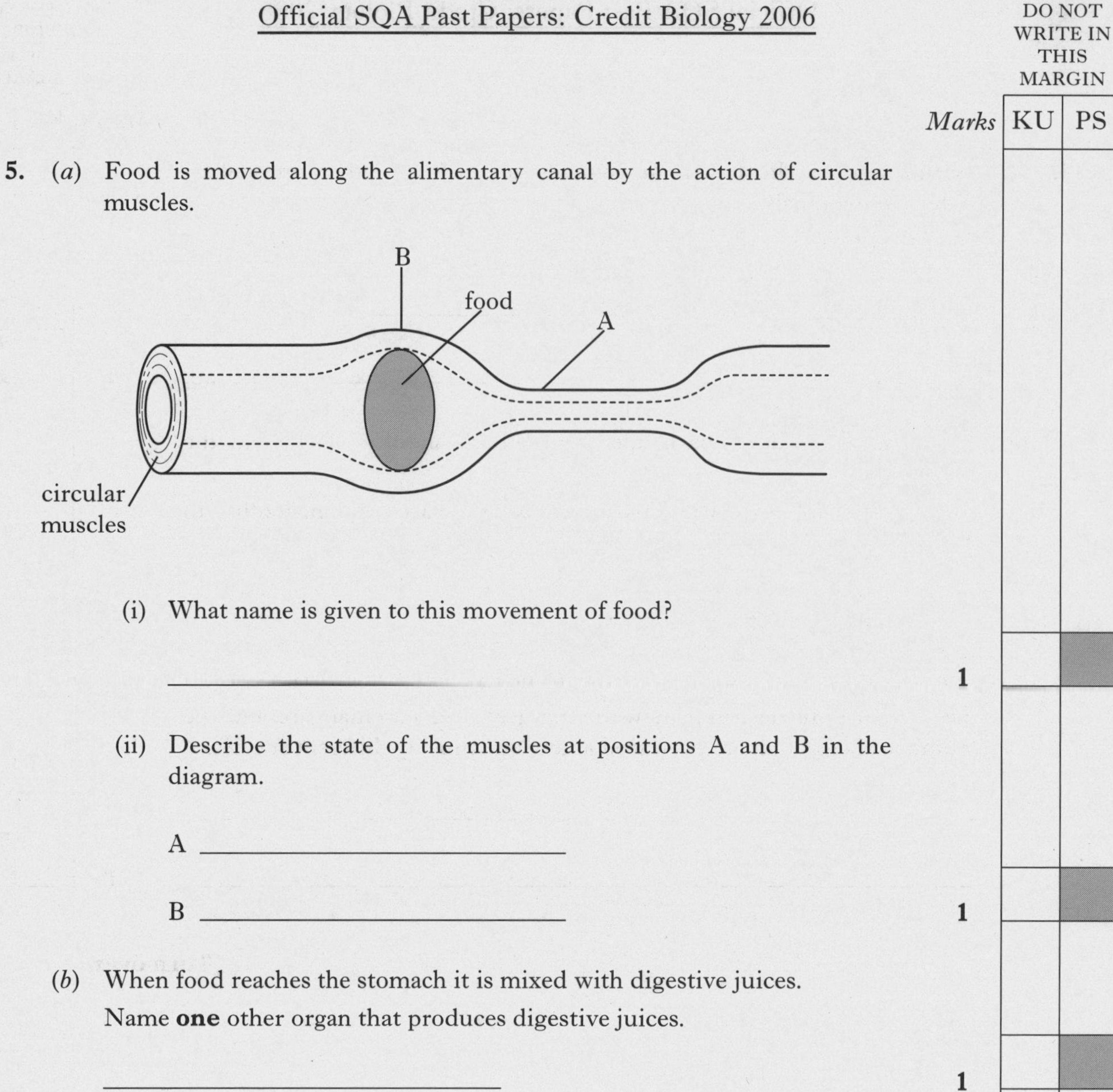

(i) What name is given to this movement of food?

________________________ 1

(ii) Describe the state of the muscles at positions A and B in the diagram.

A ________________________

B ________________________ 1

(*b*) When food reaches the stomach it is mixed with digestive juices.

Name **one** other organ that produces digestive juices.

________________________ 1

(*c*) The table shows some of the daily vitamin and mineral requirements of teenagers.

	Daily requirement (mg)				
Sex	*Vitamin B3*	*Vitamin C*	*Calcium*	*Iron*	*Zinc*
Girls	13	40	1000	15	7
Boys	17	40	800	11	10

(i) Which substance is required in equal quantities by both sexes?

________________________ 1

(ii) Which substances are required in greater quantities by boys?

________________________ 1

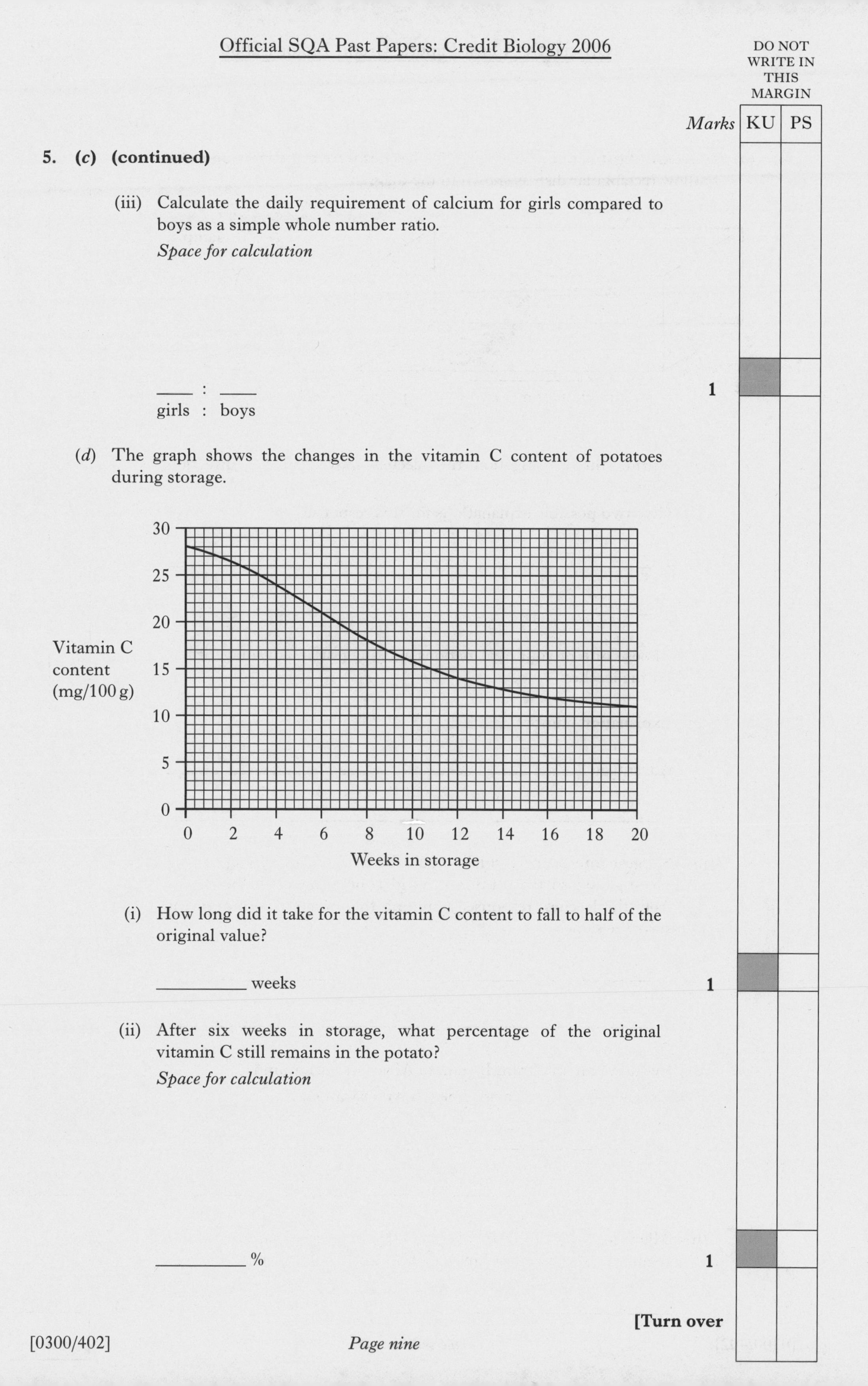

DO NOT WRITE IN THIS MARGIN

Marks KU PS

5. (*c*) (continued)

(iii) Calculate the daily requirement of calcium for girls compared to boys as a simple whole number ratio.

Space for calculation

____ : ____

girls : boys **1**

(*d*) The graph shows the changes in the vitamin C content of potatoes during storage.

(i) How long did it take for the vitamin C content to fall to half of the original value?

__________ weeks **1**

(ii) After six weeks in storage, what percentage of the original vitamin C still remains in the potato?

Space for calculation

__________ % **1**

[Turn over

DO NOT WRITE IN THIS MARGIN

Marks | KU | PS

6. (a) In an investigation into behaviour, five leeches were placed in water in a shallow rectangular dish as shown in the diagram.

(i) During the investigation the leeches moved in the direction shown.

Give **two** possible explanations for this response.

1 ______________________________

2 ______________________________ **2**

(ii) Choose **one** of your explanations and suggest an advantage it has for the leeches.

Explanation number ________

Advantage ______________________________

______________________________ **1**

(iii) Suggest **one** change which should be made to the set up of the investigation so that only one valid conclusion could be drawn from the leeches' response, assuming the direction of movement stays the same.

____

______________________________ **1**

(b) (i) Swallows migrate from Britain to Africa in the autumn.

Explain how this behaviour benefits the swallows.

____

______________________________ **1**

(ii) Migration is an example of a type of behaviour that is repeated regularly. What name is given to this type of behaviour?

______________________________ **1**

DO NOT WRITE IN THIS MARGIN

Marks KU PS

7. In an investigation, three 25 g samples of sultanas were put into separate beakers of distilled water, as shown below.

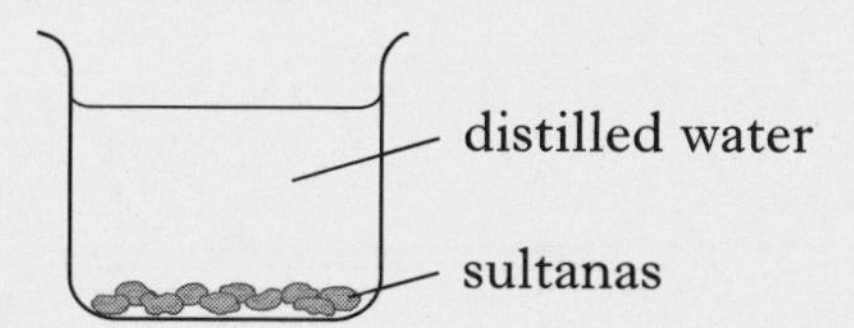

After 24 hours, the sultanas were removed from the water, blotted on filter paper and reweighed. The results are shown in the table.

Sample	*Mass after 24 hours* (g)	*Percentage change in mass*
1	32·5	30·0
2	32·2	28·8
3	32·4	

(*a*) Complete the table with the percentage change in mass of the sultanas in sample 3.

Space for calculation

1

(*b*) The change in mass of the sultanas was caused by the movement of water.

(i) Name this process.

________________________ 1

(ii) Explain the results in terms of water concentrations.

____

____ 1

(*c*) Which of the following is the best reason for blotting the sultanas before reweighing?

Tick the correct box.

☐ To stop them sticking together ☐ To remove external sugar solution

☐ To remove external water ☐ To make sure the sultanas were dried 1

DO NOT WRITE IN THIS MARGIN

Marks | KU | PS

8. The following statements refer to stages in mitosis.

1 Chromosomes become visible as pairs of chromatids.

2 Spindle fibres form.

3 ______________________________

____

4 Chromatids are pulled to opposite ends of the cell.

5 The nuclear membranes form.

6 The cytoplasm divides and two daughter cells are formed.

(*a*) Complete the sequence by writing in a description of the missing stage. **1**

(*b*) After mitosis, the daughter cells have the same number of chromosomes as the parent cell.

Explain why this is important.

____

______________________________ **1**

DO NOT WRITE IN THIS MARGIN

Marks | KU | PS

9. Read the following passage and answer the questions using information from it.

Adapted from *The Herald*, October 2003

Scientists say that the North Sea is becoming too hot for many of the fish which are included in the normal Scottish diet. Experts are blaming global warming for driving the plankton, on which the fish depend, into more northern waters. As a result, stocks of cod and salmon are in danger of collapse. At the same time, more exotic species such as red mullet, horse mackerel and black bream are increasing off the east coast of Britain.

Sand eels are also dwindling in number, and this may be having a knock-on effect on the coastal birds which feed on them. A survey of their habitats showed breeding rates for puffins, kittiwakes, guillemots and razorbills to be the lowest on record.

These trends are based on the monitoring of plankton populations. They may help to explain why a reduction in fishing has not led to a full recovery of fish populations.

Two particular episodes are blamed. The first occurred in the late 1970s and was caused by an inflow of low-temperature, low-salinity water from the North Atlantic. This was due to a high release of Arctic ice into the ocean. The second occurred in the 1980s, and this time it was an inflow of water at higher temperatures and high salinity.

(*a*) What effect is global warming having on the plankton in the North Sea?

______________________________ **1**

(*b*) Name **two** fish species which are decreasing in numbers in the North Sea.

1 ______________ 2 ______________ **1**

(*c*) Suggest a reason why exotic fish species are increasing in number off the east coast of Britain.

____

______________________________ **1**

(*d*) Explain the possible link between global warming and the expected reduction in the numbers of coastal birds.

____

______________________________ **1**

(*e*) In what **two** ways did the water which caused problems in the 1980s differ from that which caused problems in the 1970s?

1 ______________________________

2 ______________________________ **1**

Marks KU PS

10. The diagram represents the structures involved in a reflex action which occurs when a finger touches a flame.

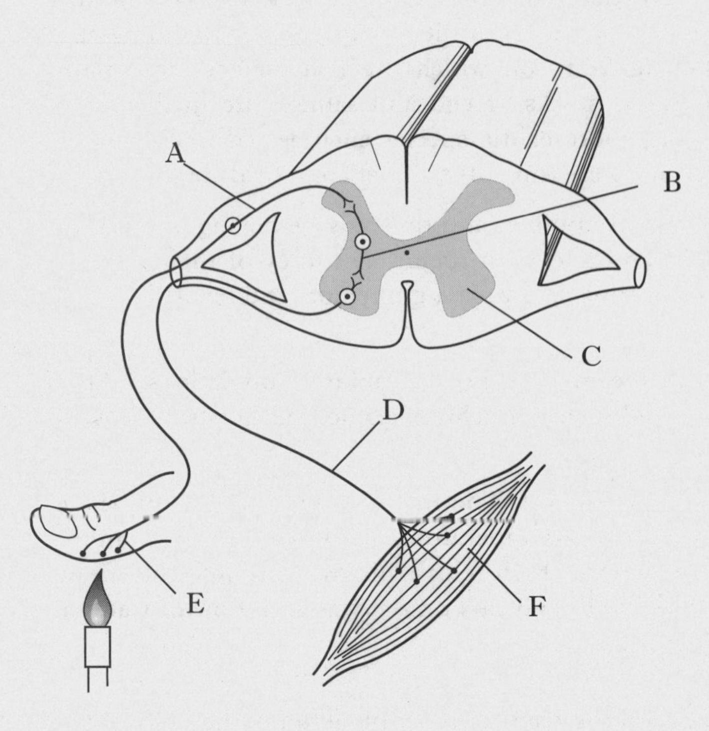

(*a*) Complete the table below with the correct letters from the diagram to identify the stages of the reflex action and with a description of the missing stage.

Stage	*Letter*
Stimulus detected by sensory receptor	
Information sent along a sensory nerve cell	
	B
Impulse sent along motor nerve cell	
Response made by effector organ	

2

DO NOT WRITE IN THIS MARGIN

Marks KU PS

10. (continued)

(*b*) An investigation was carried out on the response of the pupil of the eye. A volunteer was seated in a dark room and a torch was switched on. The diameter of the volunteer's pupil was measured.

This was repeated at different distances from the volunteer.

The results are shown on the graph below.

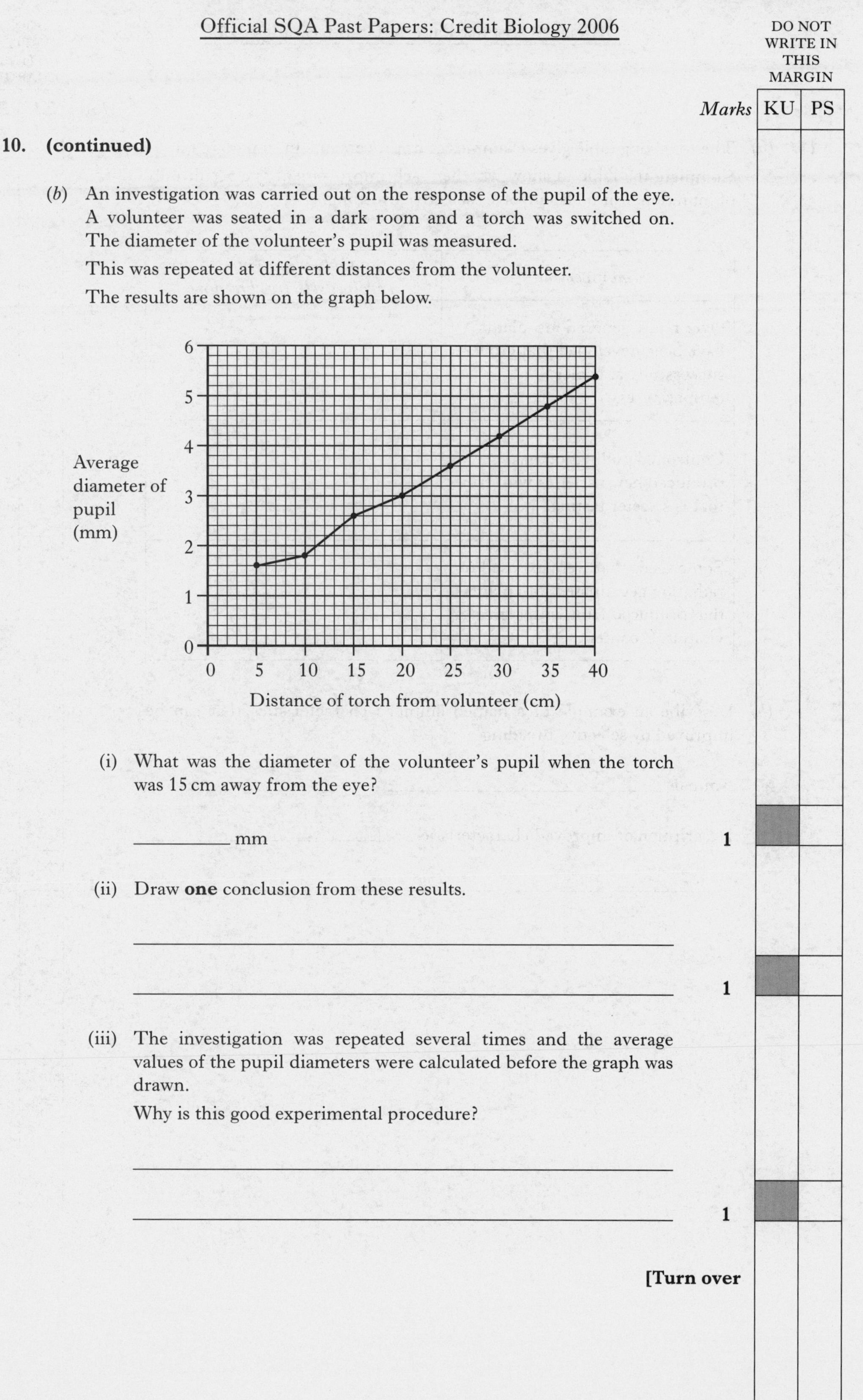

(i) What was the diameter of the volunteer's pupil when the torch was 15 cm away from the eye?

__________ mm **1**

(ii) Draw **one** conclusion from these results.

____

____ **1**

(iii) The investigation was repeated several times and the average values of the pupil diameters were calculated before the graph was drawn.

Why is this good experimental procedure?

____

____ **1**

[Turn over

DO NOT WRITE IN THIS MARGIN

Marks | KU | PS

11. (*a*) The following table gives examples of improvements in tomato plants.

Complete the table to show whether each improvement is a result only of mutation or if it also involves selective breeding.

Improvement	*Only mutation/ Involves selective breeding*
Over many generations, plants have been developed that grow successfully at cooler temperatures.	
Controlled pollination has produced new varieties with fruit that is sweeter tasting.	
Some seeds that were exposed to radiation germinated into plants that produced fruit with a greater vitamin C content.	

2

(*b*) Describe an example of a named animal's characteristics that can be improved by selective breeding.

Animal ______________________________

Description of improved characteristic ______________________________

____ **1**

DO NOT WRITE IN THIS MARGIN

Marks KU PS

12. Tay-Sachs disease is an inherited condition which affects the nerves. Different forms of the same gene determine its effect.

T (dominant) represents the normal form of the gene.
t (recessive) represents the form of the gene which causes the disease.

The family tree diagram shows a pattern of inheritance of the disease.

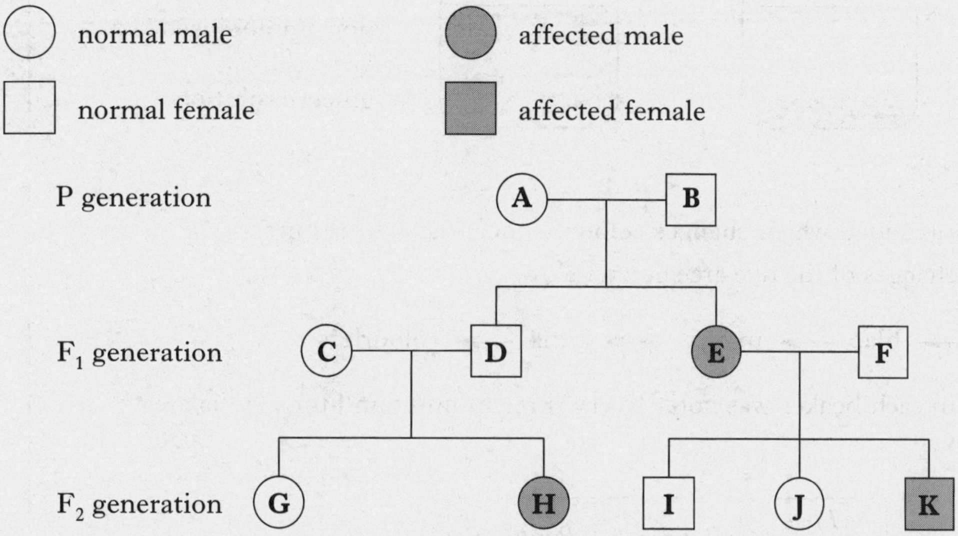

normal male — affected male

normal female — affected female

P generation: A — B

F_1 generation: C — D, E — F

F_2 generation: G, H, I, J, K

(*a*) (i) Complete the table by writing the genotypes of persons **A**, **D** and **K**.

Person	*Genotype*
A	
D	
K	

2

(ii) A carrier of the disease is someone who does not show the symptoms of the disease but can pass it to their offspring.

Give the letter of **one** person from the F_2 generation who must be a carrier of the disease.

Letter ________

1

(iii) What kind of variation is shown by Tay-Sachs disease? Explain your answer.

Variation ________________

Explanation ________________________________

____

1

(*b*) What name is given to the different forms of the same gene?

____

1

DO NOT WRITE IN THIS MARGIN

Marks | KU | PS

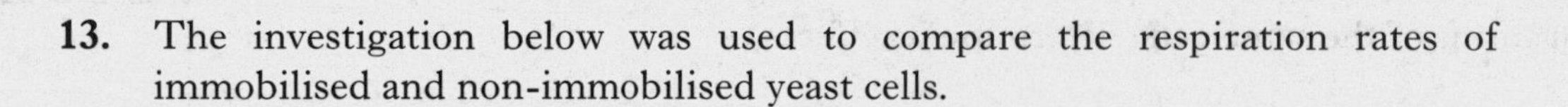

13. The investigation below was used to compare the respiration rates of immobilised and non-immobilised yeast cells.

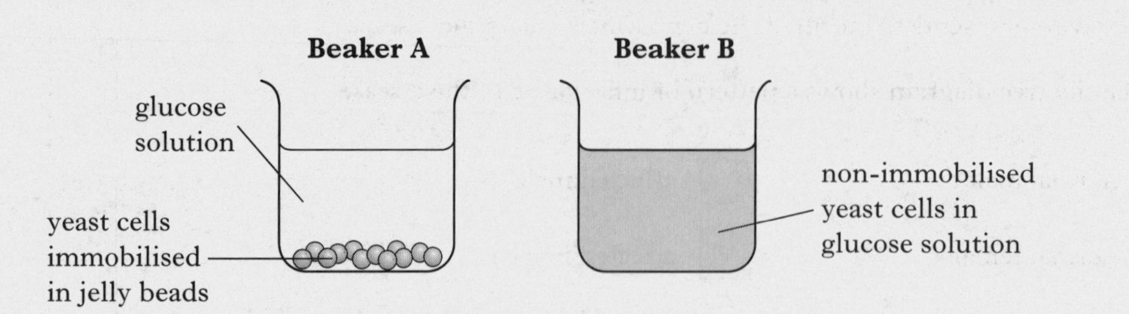

A blue dye was added which changes colour as the yeast cells respire.

The colour changes of the dye are shown below.

blue ⟶ lilac ⟶ mauve ⟶ pink ⟶ colourless

The colour in each beaker was noted every three minutes and the results are shown below.

Time (minutes)	*Beaker A*	*Beaker B*
0	blue	blue
3	blue	lilac
6	lilac	mauve
9	lilac	mauve
12	mauve	colourless
15	mauve	colourless
18	pink	colourless
21	colourless	colourless

(*a*) (i) In which beaker did the yeast cells respire faster?
Give a reason for your answer.

Beaker __________

Reason ____ **1**

(ii) Suggest a time when the dye in beaker B might have been pink.

__________ minutes **1**

DO NOT WRITE IN THIS MARGIN

Marks KU PS

13. (*a*) (continued)

(iii) Give **two** precautions that would have to be taken to ensure a valid comparison could be made between the two beakers.

1 ____

2 ____ **2**

(*b*) Immobilised cells are used in some industrial processes.

Describe **one** advantage of using immobilised cells.

____

____ **1**

(*c*) The table gives information about respiration in yeast.

Tick the boxes to show whether each statement refers to aerobic respiration, anaerobic respiration or both.

Statement	*Aerobic*	*Anaerobic*
Oxygen is used up.		
Alcohol is produced.		
Maximum energy is released.		
Carbon dioxide is produced.		

2

[Turn over

DO NOT WRITE IN THIS MARGIN

Marks KU PS

14. Cellulase is an enzyme which is produced by some soil micro-organisms. It breaks down cellulose into simple sugars. Cellulose is present in plant cell walls.

10 cm^3 samples of cellulose paste were mixed with three different liquids and left for 24 hours. The time taken for 5 cm^3 of each cellulose mixture to run through a syringe was recorded. The results are shown in the table.

Sample	*Liquid added to cellulose paste*	*Time for* 5 cm^3 *to run through* (seconds)
A	1 cm^3 cellulase solution	126
B	1 cm^3 water	375
C	1 cm^3 soil water	200

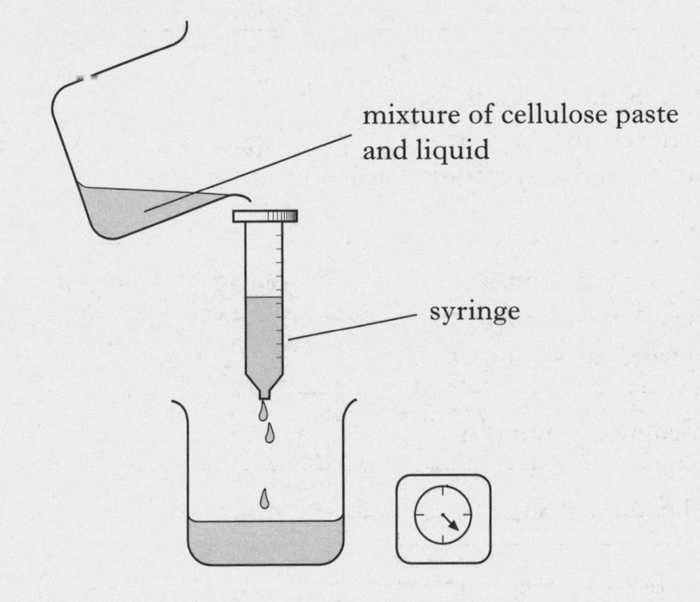

(*a*) (i) Using the results from samples A and B, describe the effect of cellulase on the thickness of cellulose paste.

______________________________ **1**

(ii) Using the results, what can be concluded about soil water?

______________________________ **1**

(*b*) (i) The samples were left in a warm place to provide optimum conditions for the enzyme.

Explain what is meant by the term *optimum conditions*.

______________________________ **1**

(ii) Cellulase enzyme is specific for cellulose.

Explain what is meant by the term *specific*.

______________________________ **1**

DO NOT WRITE IN THIS MARGIN

Marks KU PS

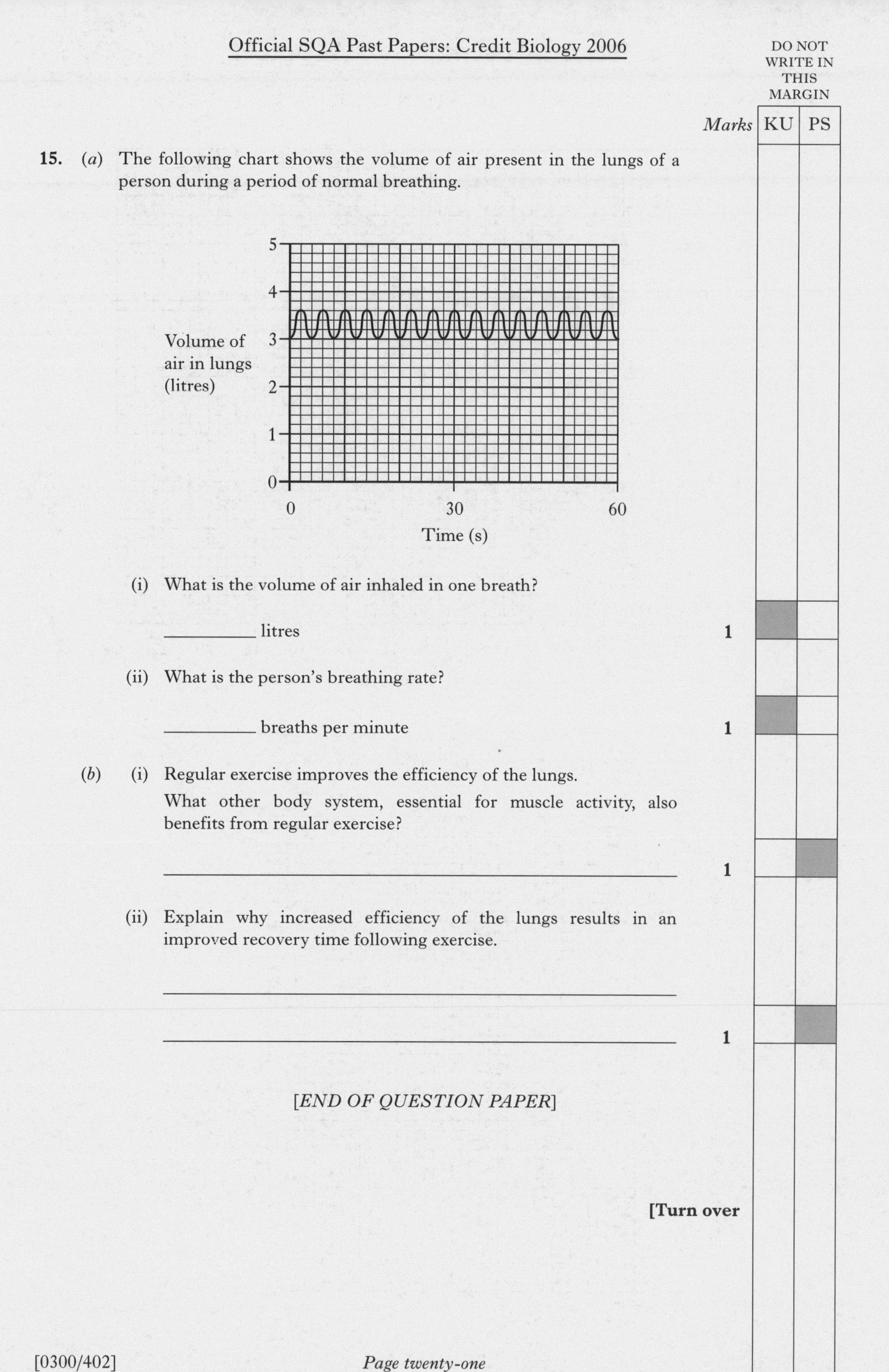

15. (*a*) The following chart shows the volume of air present in the lungs of a person during a period of normal breathing.

(i) What is the volume of air inhaled in one breath?

__________ litres **1**

(ii) What is the person's breathing rate?

__________ breaths per minute **1**

(*b*) (i) Regular exercise improves the efficiency of the lungs.

What other body system, essential for muscle activity, also benefits from regular exercise?

____ **1**

(ii) Explain why increased efficiency of the lungs results in an improved recovery time following exercise.

____

____ **1**

[*END OF QUESTION PAPER*]

[Turn over

ADDITIONAL GRAPH PAPER FOR QUESTION 1(*b*)(iii)

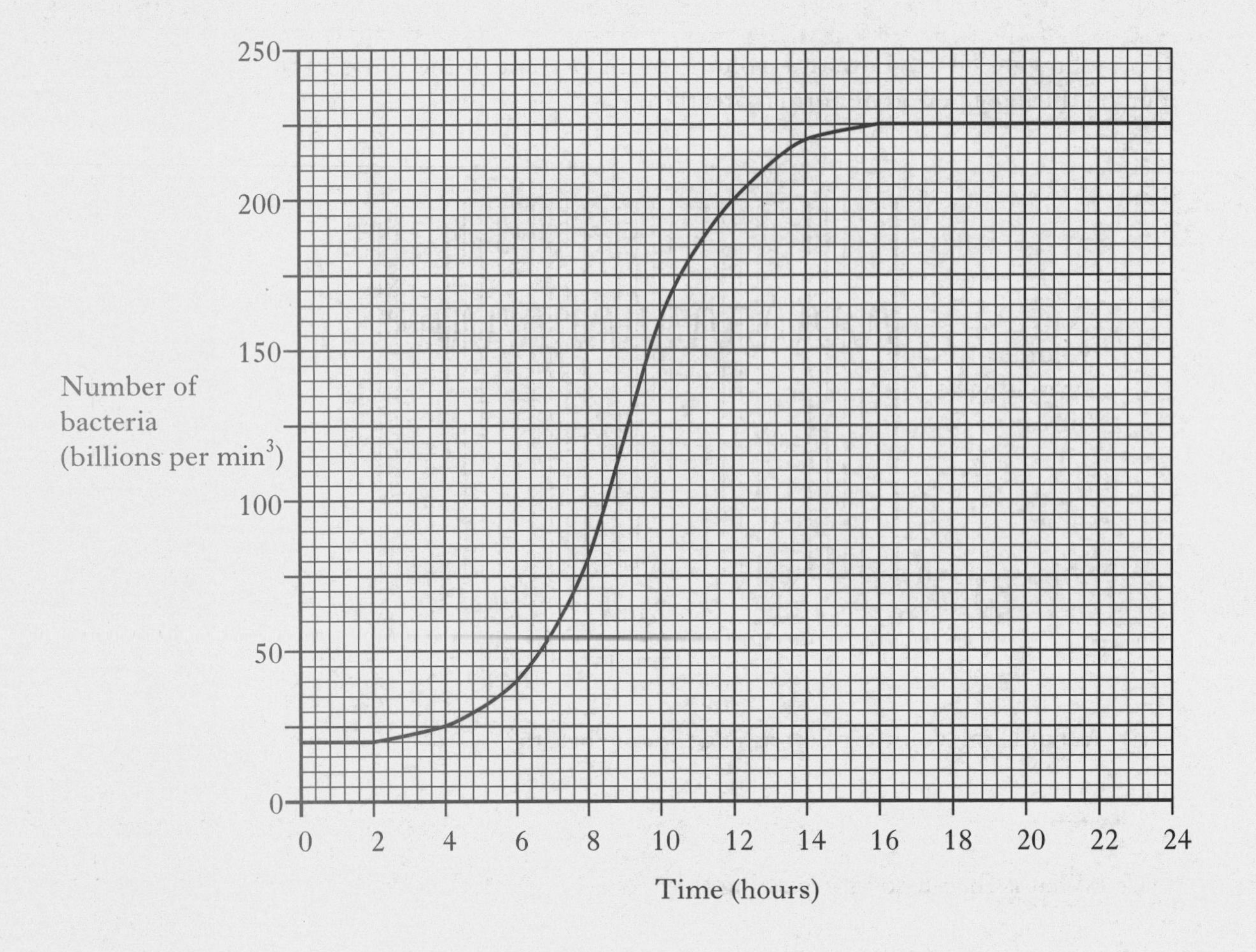

ADDITIONAL GRAPH PAPER FOR QUESTION 2(*a*)(v)

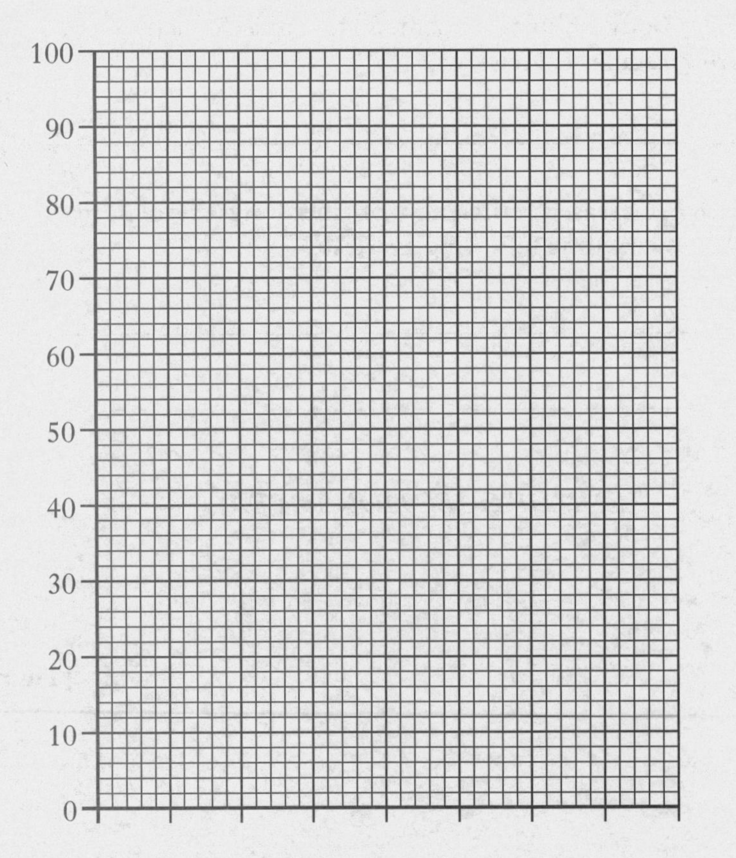

[BLANK PAGE]

FOR OFFICIAL USE

C

KU	PS

Total Marks

0300/402

NATIONAL QUALIFICATIONS 2007

MONDAY, 21 MAY
10.50 AM – 12.20 PM

BIOLOGY
STANDARD GRADE
Credit Level

Fill in these boxes and read what is printed below.

Full name of centre

Town

Forename(s)

Surname

Date of birth
Day Month Year

Scottish candidate number

Number of seat

1 All questions should be attempted.

2 The questions may be answered in any order but all answers are to be written in the spaces provided in this answer book, and must be written clearly and legibly in ink.

3 Rough work, if any should be necessary, as well as the fair copy, is to be written in this book. Additional spaces for answers and for rough work will be found at the end of the book. Rough work should be scored through when the fair copy has been written.

4 Before leaving the examination room you must give this book to the invigilator. If you do not, you may lose all the marks for this paper.

SCOTTISH QUALIFICATIONS AUTHORITY

LI 0300/402 6/31470

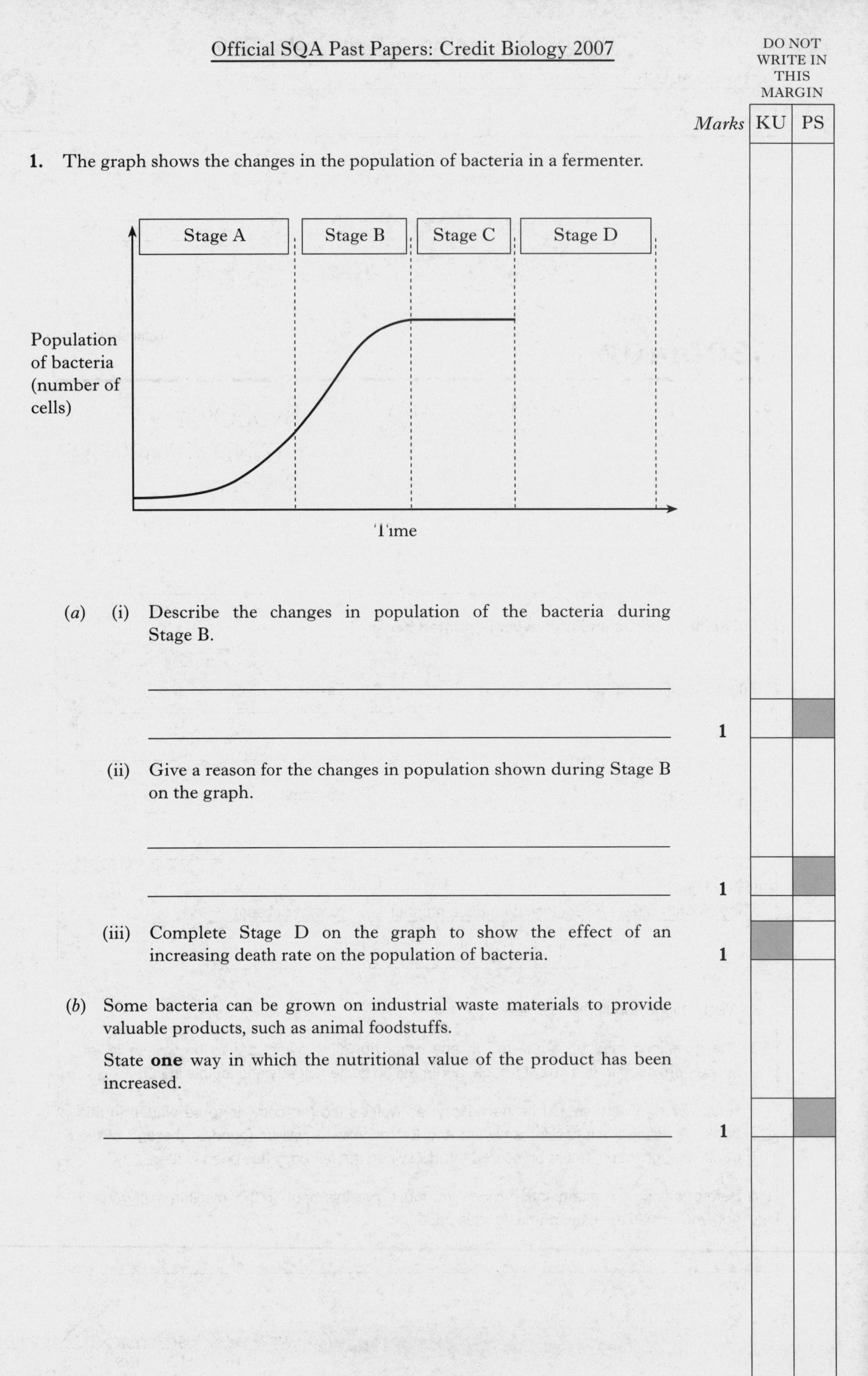

DO NOT WRITE IN THIS MARGIN

Marks KU PS

1. The graph shows the changes in the population of bacteria in a fermenter.

(*a*) (i) Describe the changes in population of the bacteria during Stage B.

____

____ **1**

(ii) Give a reason for the changes in population shown during Stage B on the graph.

____

____ **1**

(iii) Complete Stage D on the graph to show the effect of an increasing death rate on the population of bacteria. **1**

(*b*) Some bacteria can be grown on industrial waste materials to provide valuable products, such as animal foodstuffs.

State **one** way in which the nutritional value of the product has been increased.

____ **1**

DO NOT WRITE IN THIS MARGIN

Marks KU PS

2. The diagram shows some of the stages in the nitrogen cycle.

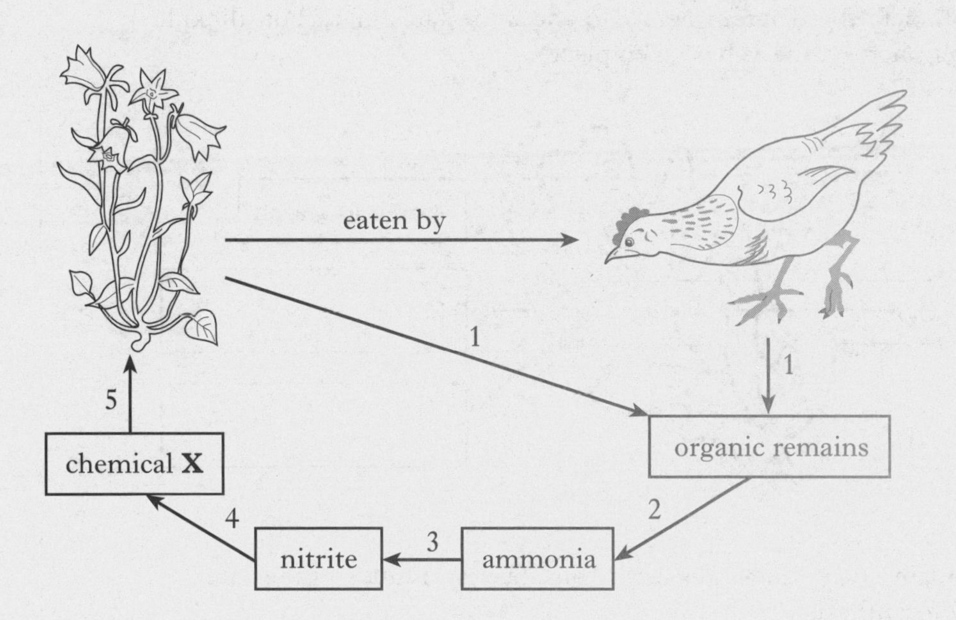

(*a*) Complete the table by giving a number from the diagram to match each of the named stages.

Stage	*Number*
Absorption	
Death	
Nitrification	
Decomposition	

2

(*b*) Name chemical **X**.

______________________________ 1

(*c*) Name the type of organism responsible for Stage 3.

______________________________ 1

[Turn over

DO NOT WRITE IN THIS MARGIN

Marks KU PS

3. (*a*) Carbon dioxide is used during photosynthesis to produce sugar.

(i) Complete the diagram below to show the fates of carbon dioxide after photosynthesis has taken place.

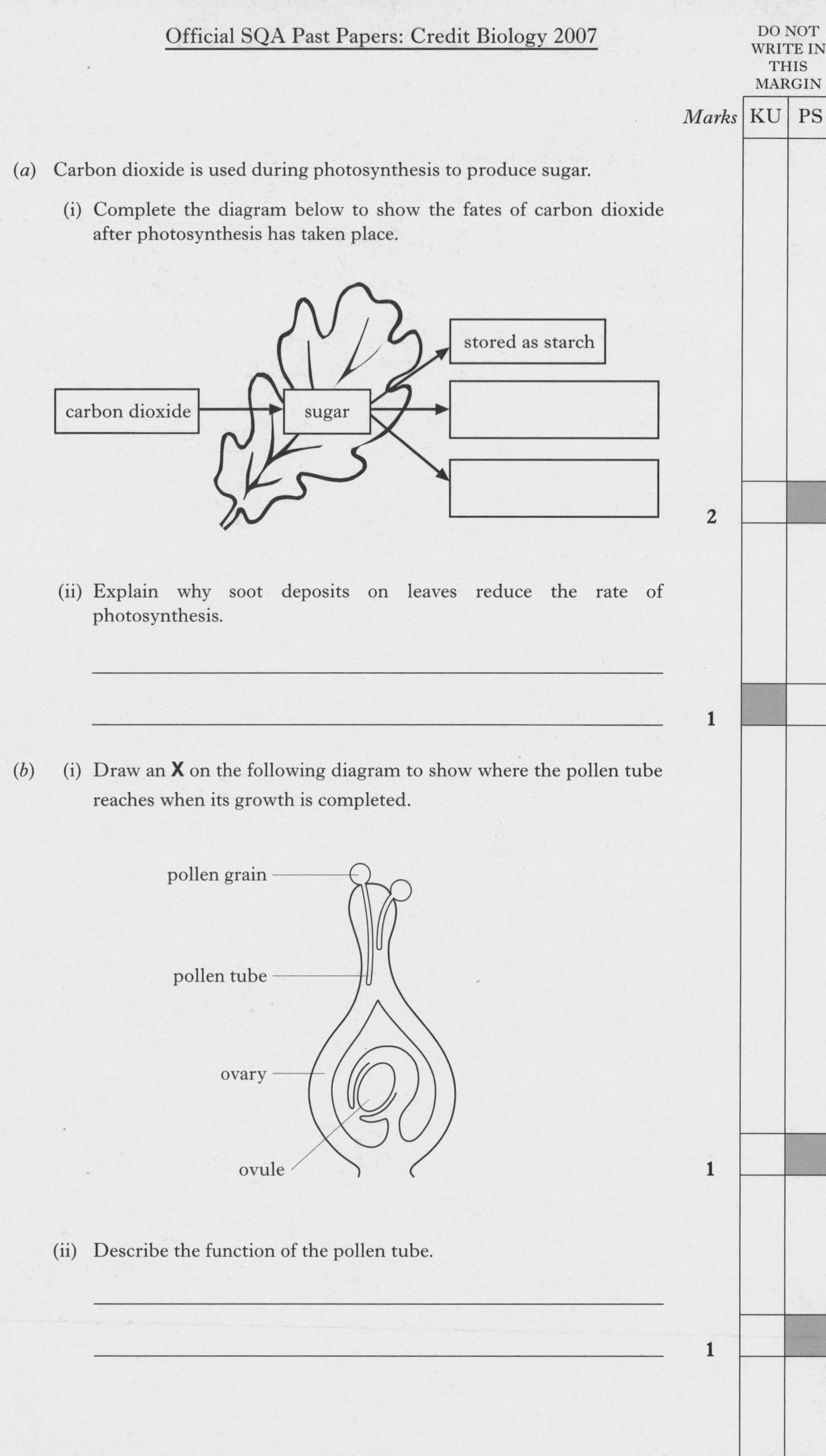

2

(ii) Explain why soot deposits on leaves reduce the rate of photosynthesis.

____

____ **1**

(*b*) (i) Draw an **X** on the following diagram to show where the pollen tube reaches when its growth is completed.

1

(ii) Describe the function of the pollen tube.

____

____ **1**

DO NOT WRITE IN THIS MARGIN

Marks | KU | PS

3. (continued)

(*c*) Tropical rain forests are being destroyed to clear land for farming. This leads to a reduction in the number of plant species.

Explain why this might lead to the extinction of some animal species.

____ **1**

(*d*) The diagrams show features of some newly discovered plants.

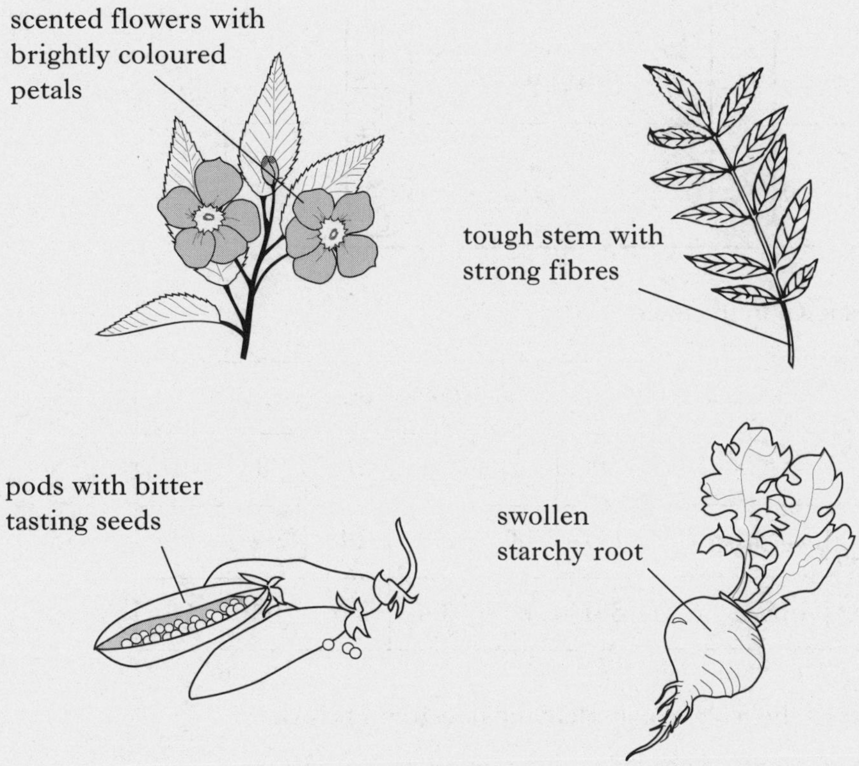

scented flowers with brightly coloured petals

tough stem with strong fibres

pods with bitter tasting seeds

swollen starchy root

Select **one** of the plant features and describe a likely use for it.

Plant feature ____________________________________

Likely use ______________________________________

____ **1**

[Turn over

DO NOT WRITE IN THIS MARGIN

Marks KU PS

4. The following investigation was set up to examine the effects of stirring on the digestion of protein.

Each piece of protein was weighed every two hours.

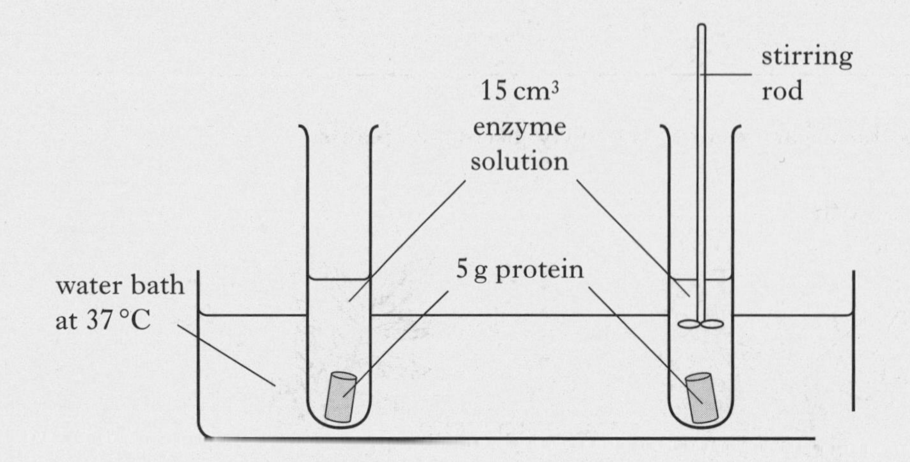

The results are shown in the table.

		Time (hours)					
		0	2	4	6	8	10
Mass of protein (g)	*not stirred*	5·0	4·7	4·3	3·8	3·2	2·5
	stirred	5·0	4·4	3·6	2·6	1·4	0·0

(*a*) Use the data in the table to complete the line graph below.

(An additional graph, if needed, will be found on page 25.)

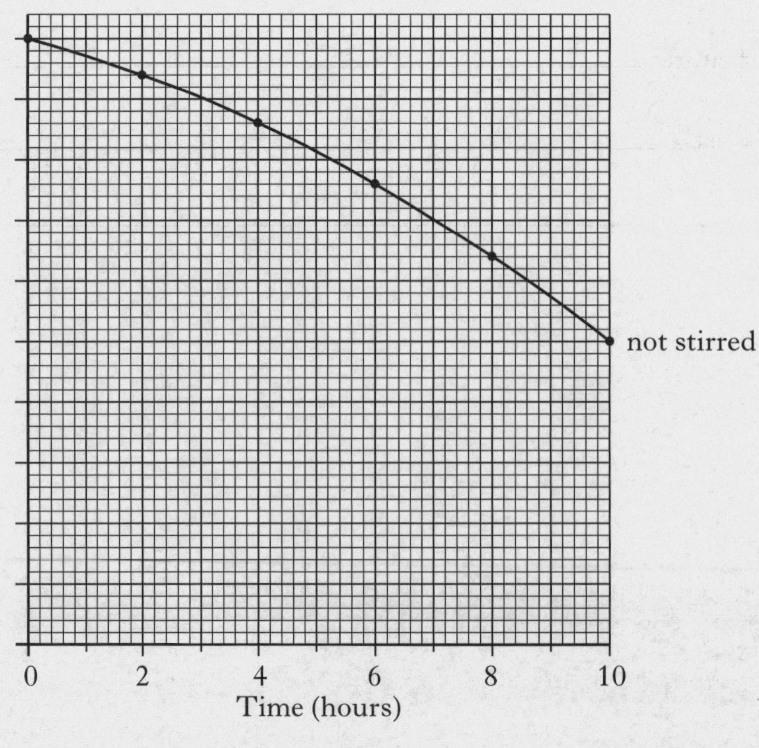

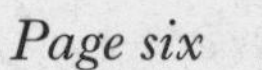

2

DO NOT WRITE IN THIS MARGIN

Marks KU PS

4. (continued)

(*b*) Which type of enzyme would produce the results shown?

________________________________ **1**

(*c*) When the protein was completely digested, no solid material remained in the tube. Explain why.

________________________________ **1**

(*d*) Name **one** factor, not already mentioned, which would need to be the same in each tube at the start of the investigation.

________________________________ **1**

(*e*) Suggest how the investigation could be improved to provide a more reliable measurement of the difference which stirring made.

________________________________ **1**

(*f*) Stirring increased the rate at which the protein was digested. Explain why this happened.

________________________________ **1**

(*g*) In the body, the stomach achieves a similar effect to stirring. Describe how this happens.

________________________________ **1**

[Turn over

DO NOT WRITE IN THIS MARGIN

Marks KU PS

5. The diagram represents a microscopic part of a kidney.

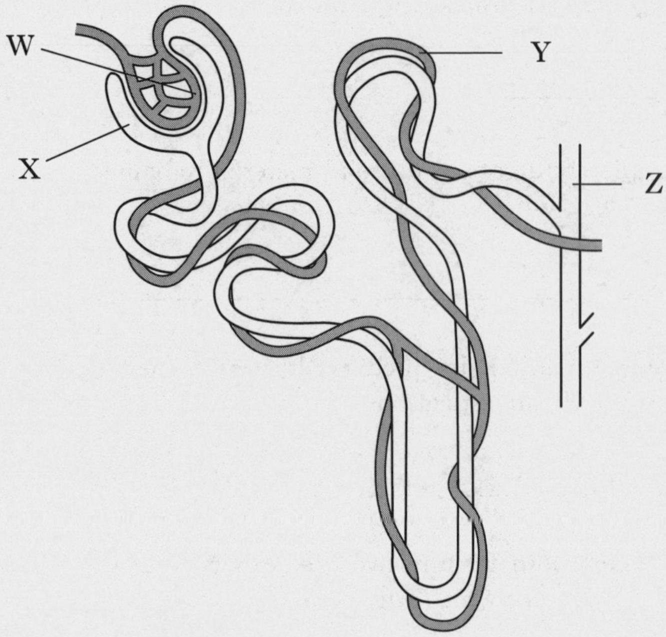

(*a*) Complete the table to show the names and functions of the structures shown on the diagram.

	Name	*Function*
W	glomerulus	
X		collection of filtrate
Y		reabsorption
Z	collecting duct	

2

DO NOT WRITE IN THIS MARGIN

5. (continued)

Marks | KU | PS

(*b*) The table shows information about kidney function.

Fluid	*Component* (g per 100cm³)				
	urea	*glucose*	*amino acids*	*salts*	*proteins*
blood plasma	0·03	0·10	0·05	0·9	8·0
glomerular filtrate	0·03	0·10	0·05	0·9	none
urine	1·75	none	none	0·90–3·60	none

(i) In which organ is urea produced and how is it transported to the kidneys?

Organ ________________

Means of transport ____ **1**

(ii) Name **one** component in the table which can pass through the wall of the glomerulus, and **one** component which cannot.

Can pass through ___________________________

Cannot pass through ________________________ **1**

(*c*) In one investigation, the kidneys of an adult male were found to filter 1254 cm³ of blood per minute. This produced 114 cm³ of filtrate per minute and 1·2 cm³ of urine per minute.

(i) Express these volumes as a simple whole number ratio.

Space for calculation

_______ : _______ : _______
blood filtrate urine **1**

(ii) Using the results of this investigation and information from the table, calculate the mass of urea which would be excreted by this person in 24 hours.

Space for calculation

___________ g **1**

[Turn over

DO NOT WRITE IN THIS MARGIN

Marks KU PS

6. The brown shrimp is found all round our coastline.

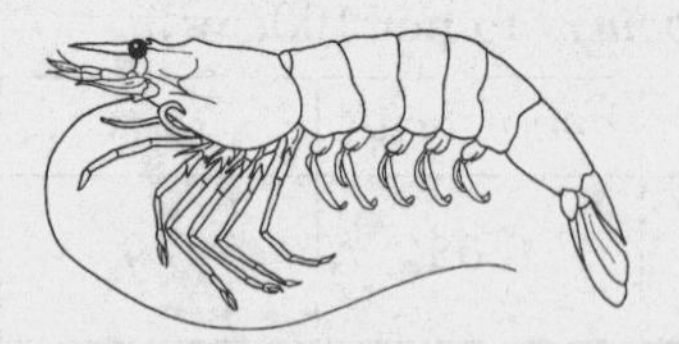

The graph shows shrimp activity and changes in their environment over a 48 hour period.

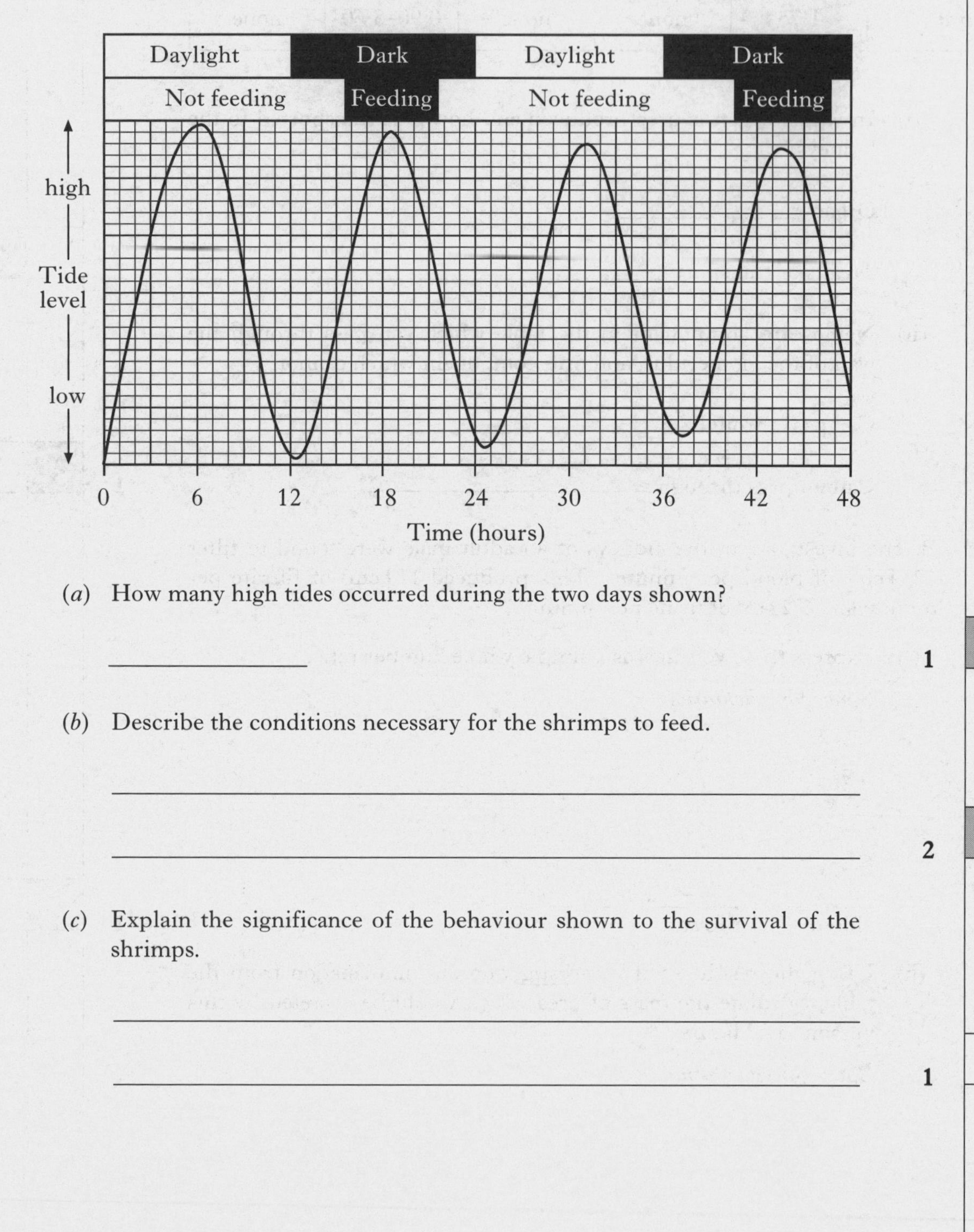

(*a*) How many high tides occurred during the two days shown?

________________ **1**

(*b*) Describe the conditions necessary for the shrimps to feed.

____

____ **2**

(*c*) Explain the significance of the behaviour shown to the survival of the shrimps.

____

____ **1**

DO NOT WRITE IN THIS MARGIN

Marks | KU | PS

7. A flower petal was examined under the microscope and then placed in a concentrated salt solution for 30 minutes. It was then re-examined under the microscope.

The diagrams show a cell from the petal before and after being in the solution.

before after

(*a*) (i) The movement of water caused the change in the appearance of the cell. What name is given to this movement of water?

________________________ **1**

(ii) In terms of water concentration, explain **why** this movement of water took place.

____

____ **1**

(*b*) Name **one** substance, other than water, which must be able to pass into a cell for its survival.

________________________ **1**

(*c*) The diagram below shows a group of cells as seen under a microscope. The field of view was 2 mm in diameter.

Calculate the average length and width of the cells.

Space for calculation

Average length ______ mm

Average width ______ mm **1**

[Turn over

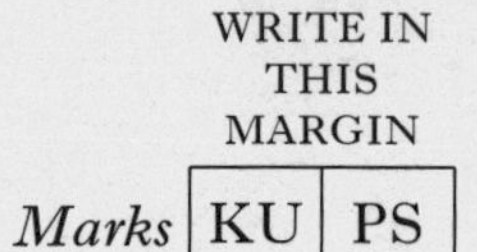

Marks KU PS

8. (*a*) The diagram shows a method used to investigate the energy content of a variety of foods.

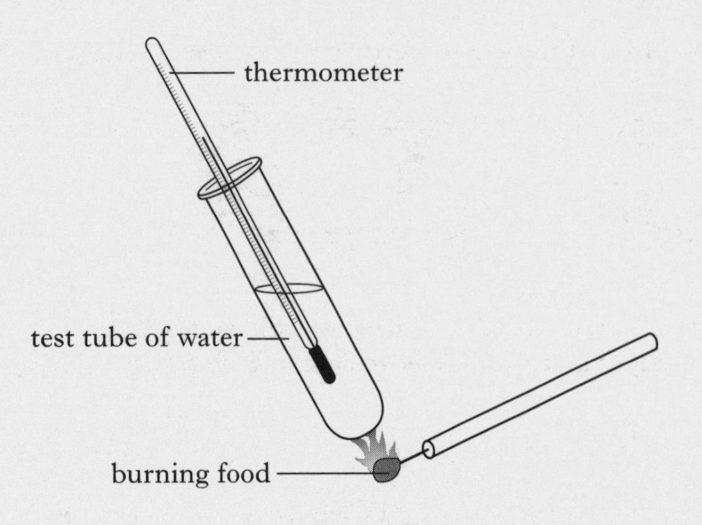

The rise in temperature can be used to calculate the energy content of each food in kilojoules.

The results are shown in the table.

Type of food	*mass* (g)	*energy content* (kilojoules)
cheese	1·0	17·0
fish	1·0	0·5
steak	1·0	13·9
carrot	1·0	1·8
apple	1·0	2·5

(i) State **two** factors, not already mentioned, that should be kept constant for a valid comparison to be made between the foods.

1 ____________________

2 ____________________ **2**

(ii) Suggest why the energy contents found in the investigation might not have been as high as expected.

____

____________________ **1**

DO NOT WRITE IN THIS MARGIN

Marks KU PS

8. (*a*) (continued)

(iii) The energy content of each food was calculated using the following formula.

Energy content (kilojoules) = temperature rise × 0·21

Calculate the energy content of 1g of chicken, if it raised the temperature of the water by 30 °C.

Space for calculation

______ kilojoules per gram **1**

(*b*) Give **one** reason, other than providing heat, why cells need energy from food.

______________________________ **1**

(*c*) Which component of food provides most energy per gram?

______________________________ **1**

[Turn over

DO NOT WRITE IN THIS MARGIN

Marks | KU | PS

9. The diagram below shows a cross-section through a joint.

X

Y

syringe

bone

(*a*) Name and describe the functions of parts X and Y on the diagram.

Part X Name ________________

Function ____

____ **1**

Part Y Name ________________

Function ____

____ **1**

(*b*) Some of the synovial fluid from inside a joint can be removed for medical tests using a syringe as shown in the diagram.

(i) Name the part of the joint which produces the synovial fluid and describe the function of the fluid.

Produced by ____

Function ____ **1**

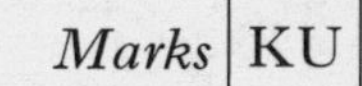

Marks | KU | PS

9. (*b*) (continued)

(ii) The table below describes the features of the fluid which lead to the diagnosis of several joint abnormalities.

		Feature of synovial fluid		
		Viscosity	*Cloudiness*	*Colour*
Diagnosis	*Normal*	high	zero	light yellow
	Inflammation	low	slight	dark yellow
	Infection	low	high	dark yellow
	Blood leakage	intermediate	high	pink

Use the information from the table to complete the paired statement key to identify the diagnoses.

1. Fluid pink .... Blood leakage

   Fluid not pink .... go to 2

2. Low viscosity .... [ ]

   High viscosity .... [ ]

3. [ ] .... Infection

   [ ] .... [ ]

2

[Turn over

DO NOT WRITE IN THIS MARGIN

Marks KU PS

10. Read the following passage and answer the questions based on it.

Invasion of the Chinese Mitten Crab

Adapted from *Biological Sciences Review*, Volume 15, Number 2.

The Chinese mitten crab, *Eriocheir sinensis*, lives in fresh water as an adult, but it breeds in the lower reaches of estuaries and spends part of its early life in seawater.

It looks different from other crabs. Its claws are covered in a coating of fine brown hairs resembling mittens. This type of crab is a problem because it burrows into river banks, causing them to collapse and silt up river channels.

The mitten crab is not native to Europe. They were recorded in the River Thames in the 1930s. Their larvae may have been transported to the river in ships' ballast water and released during dumping of this water before the ship took on cargo.

Adult mitten crabs have been known to travel thousands of kilometres in freshwater at up to 18 km per day. The young crabs, when migrating upriver, seem to be mainly herbivorous. As they grow, they become omnivorous, eating vegetation, crustaceans, insects and dead fish—in fact anything they can get a hold of! Not only is this a problem for the plants and animals that they are eating, but also they compete with native species, such as freshwater crayfish, for food.

(*a*) How does the appearance of the Chinese mitten crab differ from other crabs?

____ **1**

(*b*) State the type of environment the Chinese mitten crabs are found in at each of the following stages in their life.

(i) Early years ______________________________

(ii) Breeding times ______________________________

(iii) Mature adults ______________________________ **1**

(*c*) How is it thought that the Chinese mitten crabs arrived in Britain?

____ **1**

DO NOT WRITE IN THIS MARGIN

Marks KU PS

10. (continued)

(*d*) Describe **one** problem the Chinese mitten crab causes to the habitat and **one** problem it causes to the native community.

Habitat ______________________________ **1**

Community ______________________________ **1**

(*e*) Describe the changes in its diet as a young adult mitten crab grows.

____

______________________________ **1**

(*f*) When moving at their maximum speed, how long would it take an adult mitten crab to travel the whole length of a 45 km river?

Space for calculation

__________ days **1**

[Turn over

DO NOT WRITE IN THIS MARGIN

Marks KU PS

11. (*a*) Lactic acid is a waste product from one type of respiration. What type of respiration produces lactic acid?

______________________________ **1**

(*b*) The lactic acid content of the blood of a professional cyclist was measured while cycling at different speeds.

The graph shows the results of these measurements taken at the start of the racing season and at the end.

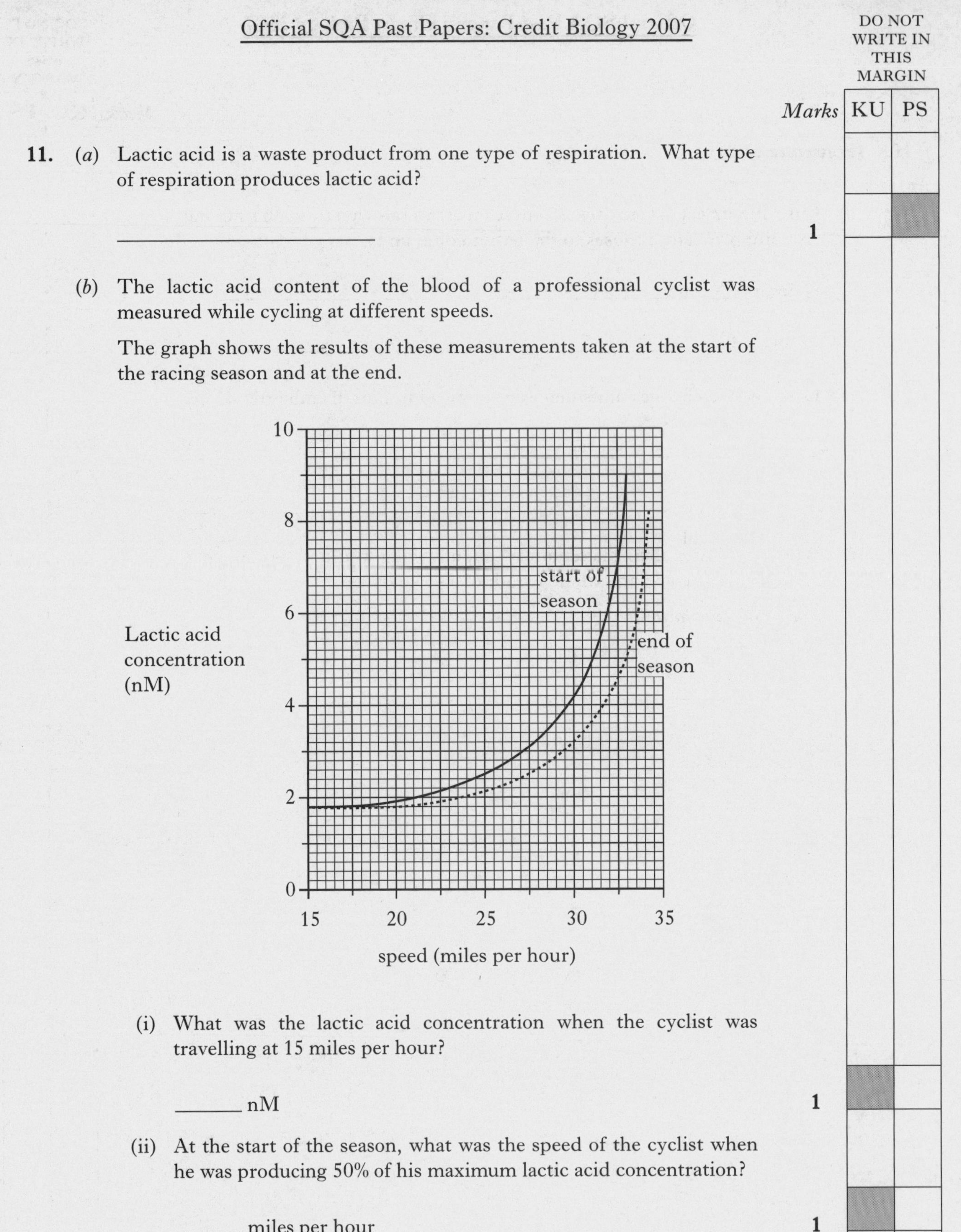

(i) What was the lactic acid concentration when the cyclist was travelling at 15 miles per hour?

_______ nM **1**

(ii) At the start of the season, what was the speed of the cyclist when he was producing 50% of his maximum lactic acid concentration?

_______ miles per hour **1**

DO NOT WRITE IN THIS MARGIN

Marks | KU | PS

11. (*b*) (continued)

(iii) When lactic acid concentration rises above 2·5 nM, the leg muscles quickly lose power and become painful.

1 What name is given to this condition?

____________________________ **1**

2 What is the maximum speed this cyclist could maintain at the start of the season?

______ miles per hour **1**

(iv) The graph shows that training improves the efficiency of muscles. Other than muscle, name **two** organs whose efficiency is improved by training.

1 __________________________

2 __________________________ **1**

[Turn over

DO NOT WRITE IN THIS MARGIN

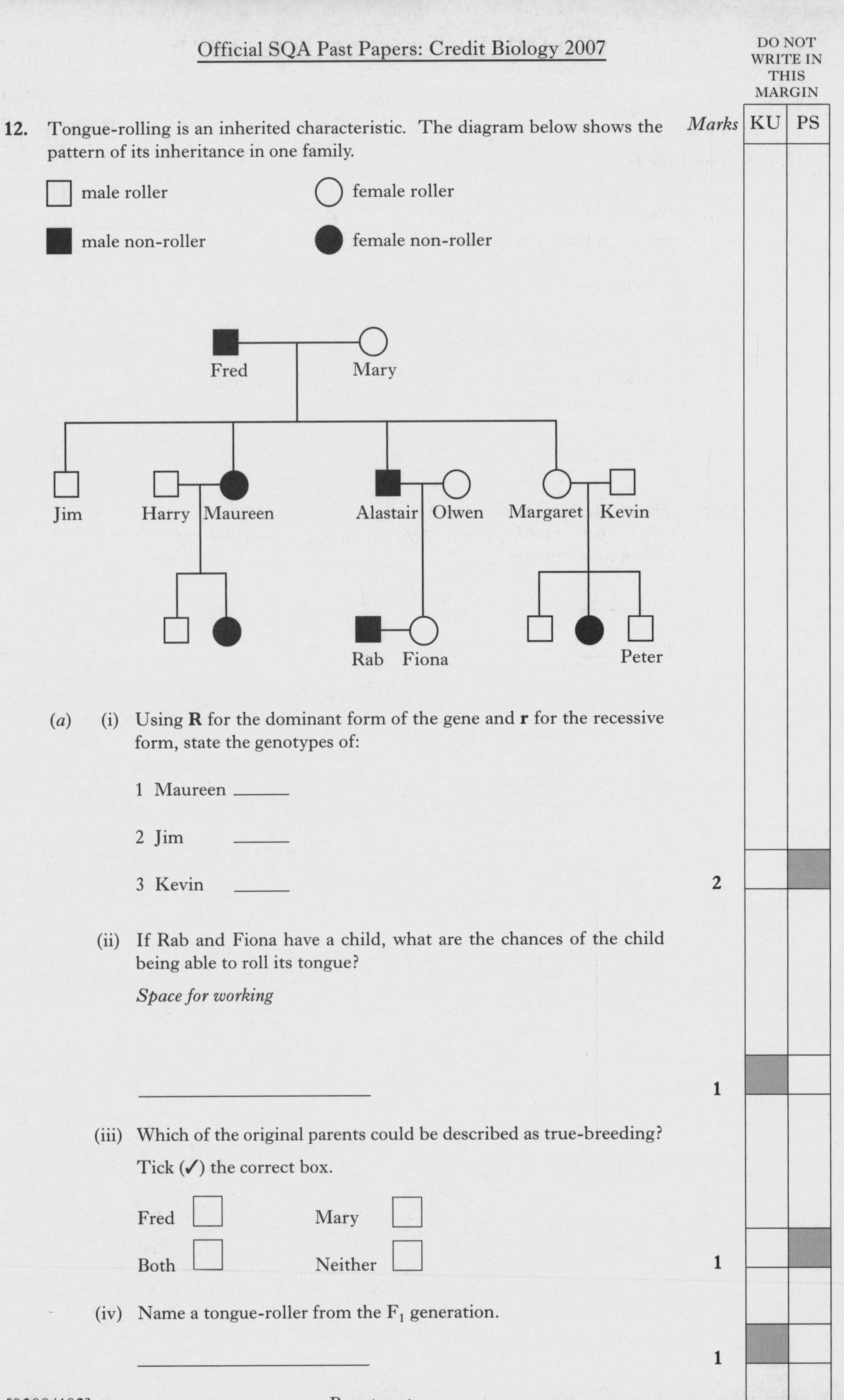

Marks	KU	PS

12. Tongue-rolling is an inherited characteristic. The diagram below shows the pattern of its inheritance in one family.

(*a*) (i) Using **R** for the dominant form of the gene and **r** for the recessive form, state the genotypes of:

1 Maureen ______

2 Jim ______

3 Kevin ______ **2**

(ii) If Rab and Fiona have a child, what are the chances of the child being able to roll its tongue?

Space for working

______________________ **1**

(iii) Which of the original parents could be described as true-breeding?

Tick (✓) the correct box.

Fred ☐ Mary ☐

Both ☐ Neither ☐ **1**

(iv) Name a tongue-roller from the F_1 generation.

______________________ **1**

DO NOT WRITE IN THIS MARGIN

12. (continued) *Marks* KU PS

(*b*) Explain why the proportions of the offspring phenotypes from genetic crosses are not always exactly as predicted.

____

____ **1**

(*c*) What term is used for the different forms of the same gene?

____ **1**

[Turn over

DO NOT WRITE IN THIS MARGIN

Marks | KU | PS

13. The diagram shows an industrial fermenter. It is fitted with a number of taps which allow substances to be added or removed.

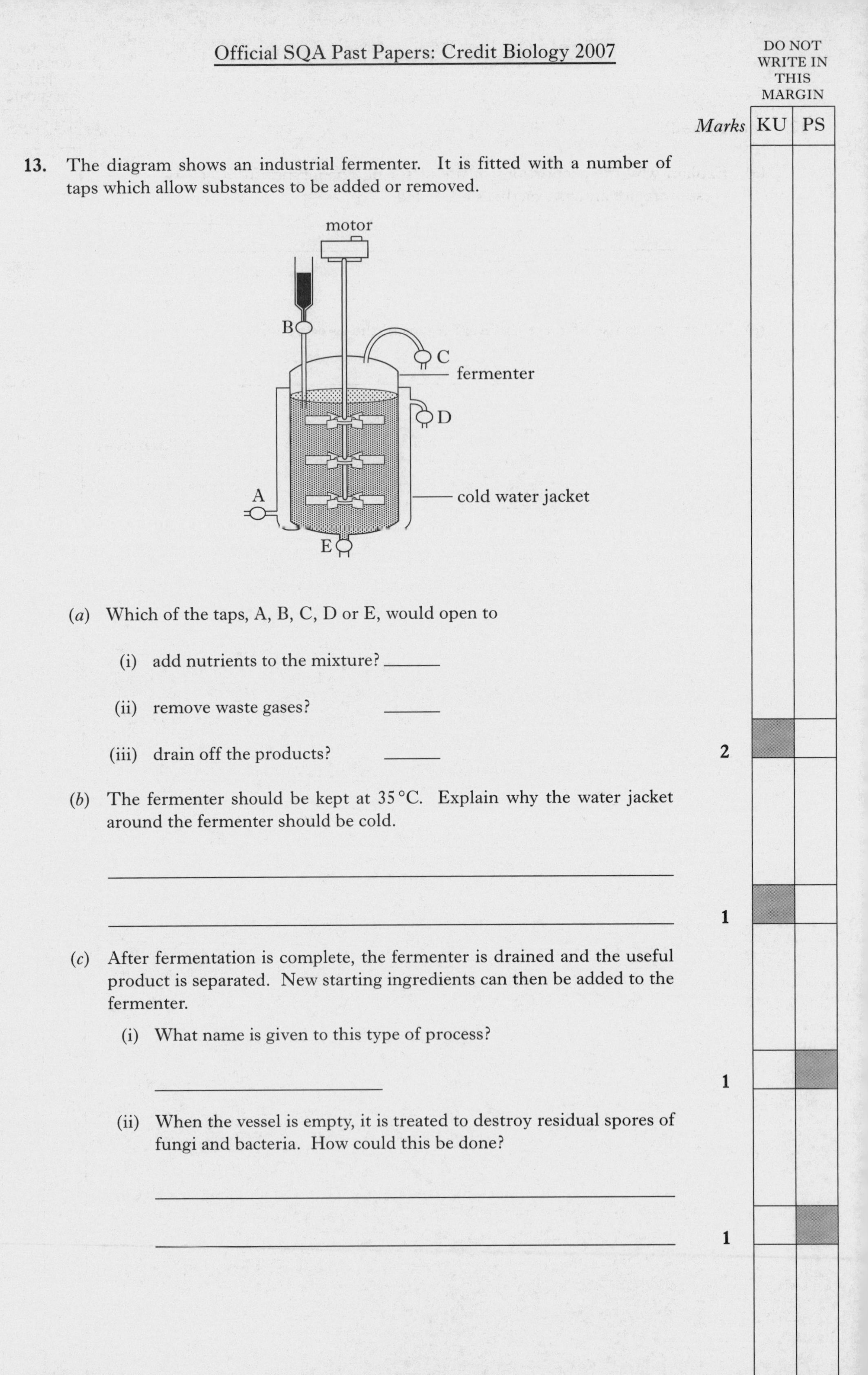

(*a*) Which of the taps, A, B, C, D or E, would open to

(i) add nutrients to the mixture? ______

(ii) remove waste gases? ______

(iii) drain off the products? ______ **2**

(*b*) The fermenter should be kept at 35 °C. Explain why the water jacket around the fermenter should be cold.

____

____ **1**

(*c*) After fermentation is complete, the fermenter is drained and the useful product is separated. New starting ingredients can then be added to the fermenter.

(i) What name is given to this type of process?

______________________ **1**

(ii) When the vessel is empty, it is treated to destroy residual spores of fungi and bacteria. How could this be done?

____

____ **1**

DO NOT WRITE IN THIS MARGIN

Marks KU PS

13. (continued)

(*d*) Barley malt extract, water and yeast were placed in the fermenter and left for several days.

The rate of fermentation was measured and the results are shown in the graph below.

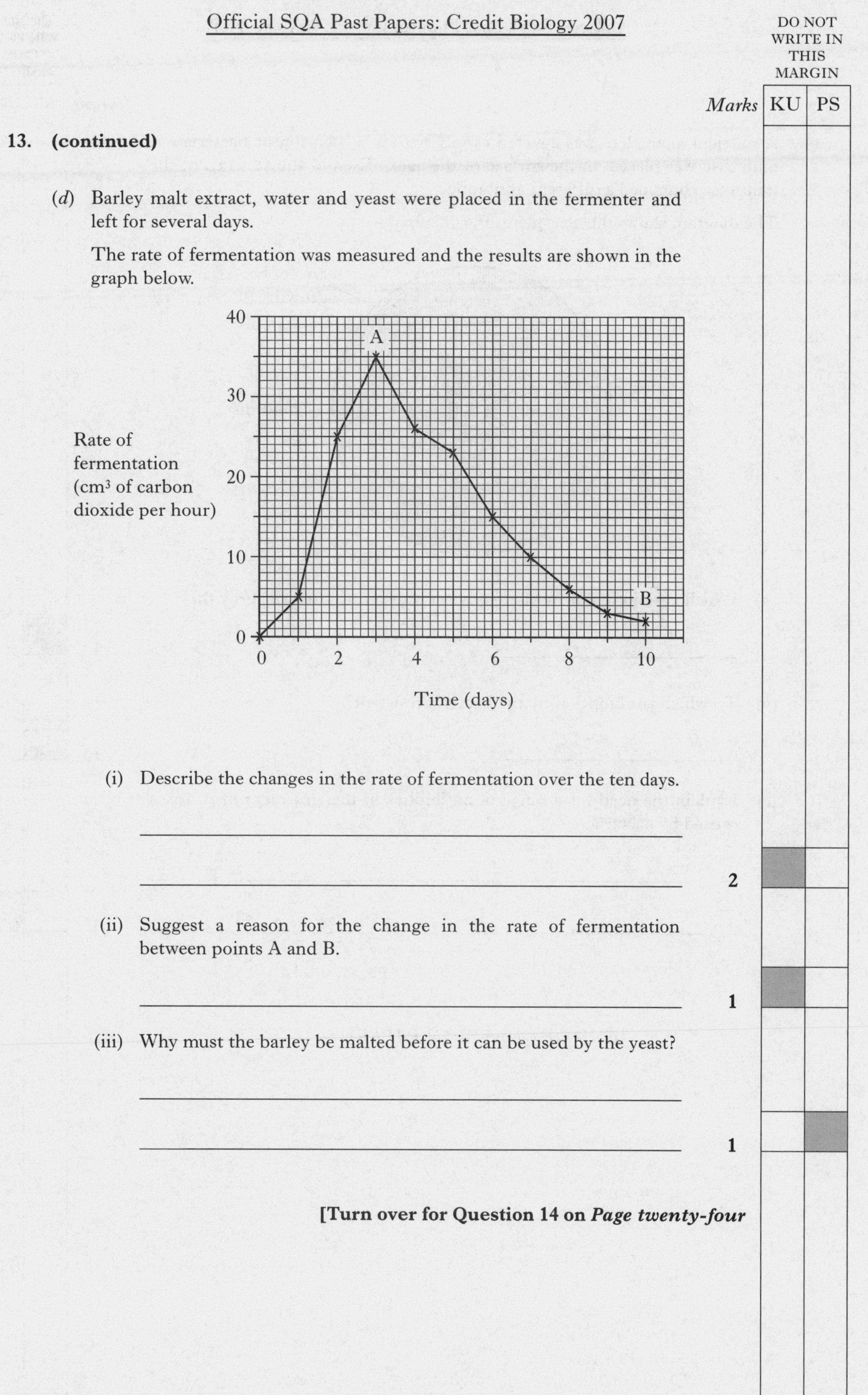

(i) Describe the changes in the rate of fermentation over the ten days.

____

____ **2**

(ii) Suggest a reason for the change in the rate of fermentation between points A and B.

____ **1**

(iii) Why must the barley be malted before it can be used by the yeast?

____

____ **1**

[Turn over for Question 14 on *Page twenty-four*

DO NOT WRITE IN THIS MARGIN

Marks | KU | PS

14. A nutrient agar plate was covered evenly with a suspension of bacteria. A multidisc was placed on the surface of the agar. Each of the six ends of the multidisc contained a different antibiotic.

The diagram shows the agar plate after incubation.

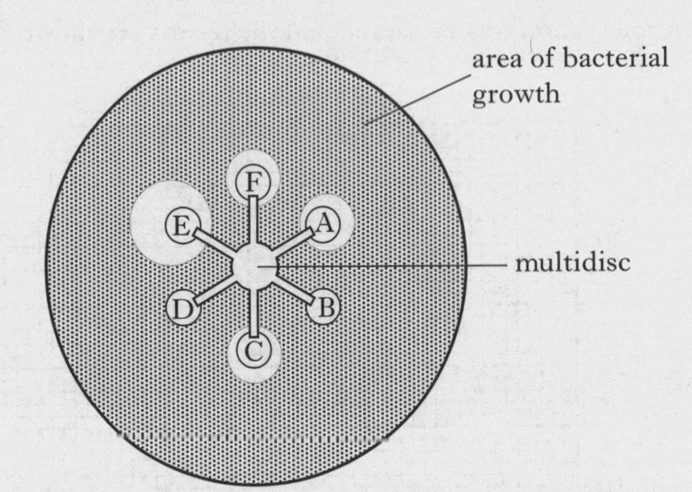

(*a*) Which antibiotic was most effective at preventing bacterial growth?

____________________ **1**

(*b*) To which antibiotics were the bacteria resistant?

____________________ **1**

(*c*) Explain the need for a range of antibiotics in the treatment of diseases caused by bacteria.

____

____ **1**

[*END OF QUESTION PAPER*]

ADDITIONAL GRAPH PAPER FOR QUESTION 4(*a*)

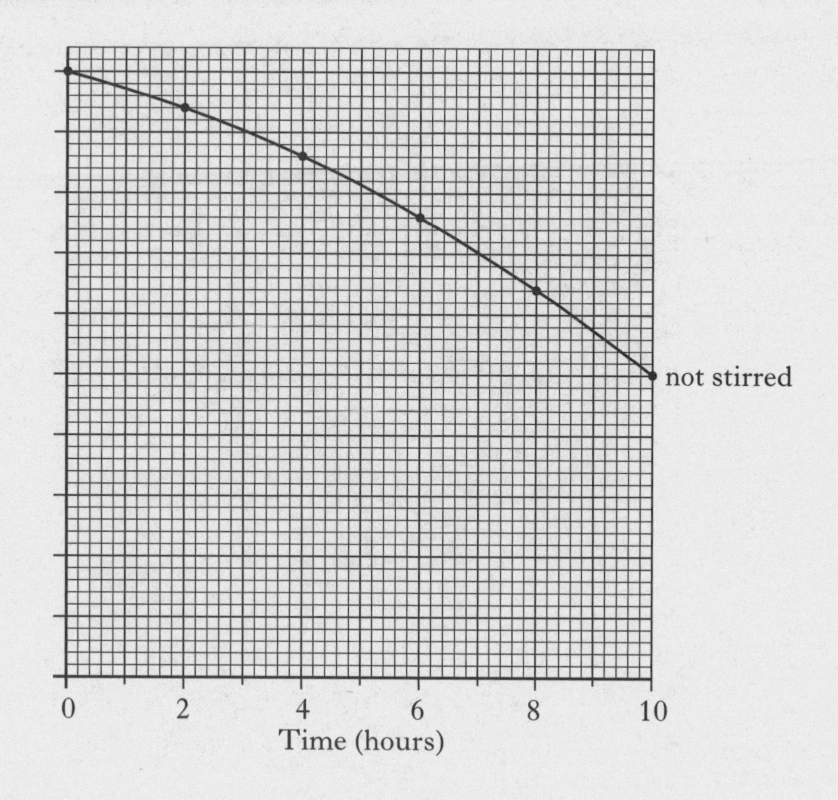

[Turn over

SPACE FOR ANSWERS
AND FOR ROUGH WORKING

2008 | Credit

[BLANK PAGE]

C

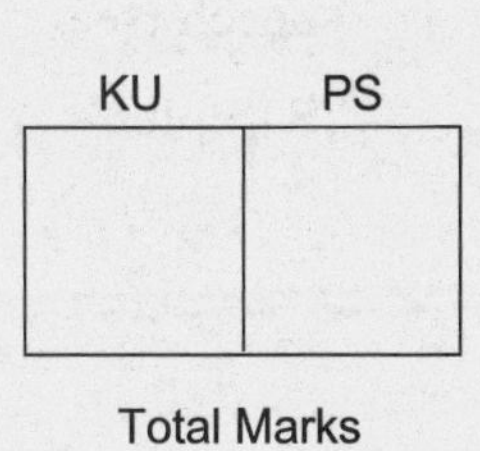

0300/402

NATIONAL QUALIFICATIONS 2008

TUESDAY, 27 MAY
10.50 AM – 12.20 PM

BIOLOGY
STANDARD GRADE
Credit Level

Fill in these boxes and read what is printed below.

Full name of centre

Town

Forename(s)

Surname

Date of birth
Day Month Year

Scottish candidate number

Number of seat

1 All questions should be attempted.

2 The questions may be answered in any order but all answers are to be written in the spaces provided in this answer book, and must be written clearly and legibly in ink.

3 Rough work, if any should be necessary, as well as the fair copy, is to be written in this book. Additional spaces for answers and for rough work will be found at the end of the book. Rough work should be scored through when the fair copy has been written.

4 Before leaving the examination room you must give this book to the invigilator. If you do not, you may lose all the marks for this paper.

SQA

LI 0300/402 6/30670

DO NOT WRITE IN THIS MARGIN

Marks KU PS

1. (*a*) A comparison was made between the types of invertebrate animals living on the branches and leaves on an oak tree with those living on a beech tree.

Samples were collected as shown below.

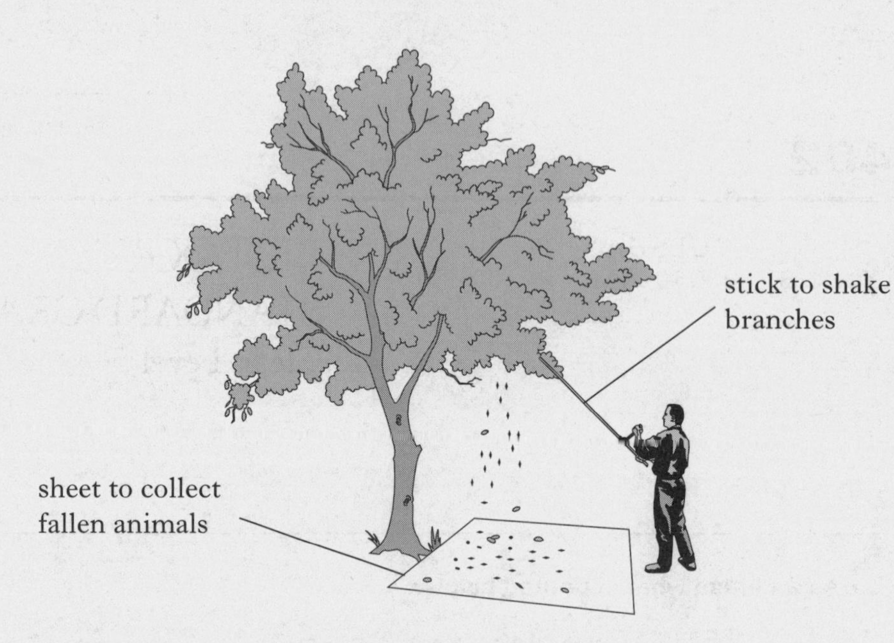

(i) Give **two** variables which should be kept constant to make the comparison valid when using this technique.

1 ______________________________

2 ______________________________ **1**

(ii) The samples collected were not representative of all the invertebrates living on the trees. Suggest a reason for this.

______________________________ **1**

(iii) Measurement of abiotic factors such as light intensity may be recorded at the same time as sampling. Identify a possible source of error for a **named** measurement technique and explain how it might be minimised.

Measurement technique ______________________________

Source of error ______________________________

How to minimise it ______________________________

______________________________ **1**

1. (continued)

(*b*) An investigation was carried out into the effect of light intensity on the distribution of a plant species. At eight different measurement points in a garden, the average light intensity was measured and the percentage ground cover of the plant was recorded.

The results are shown below.

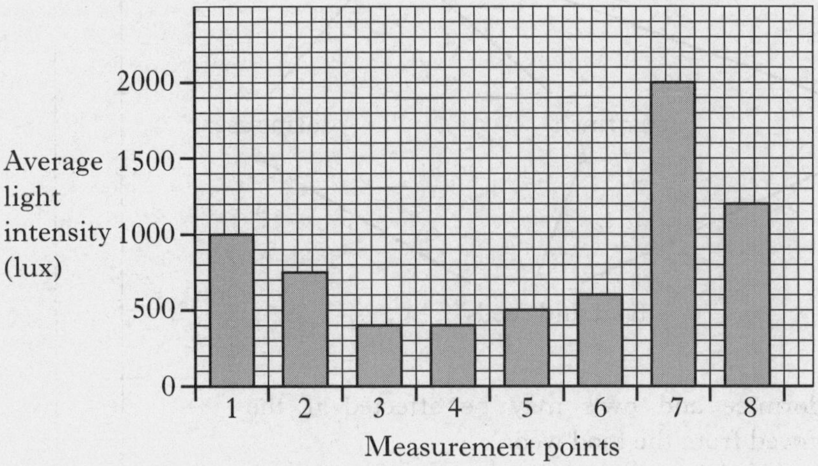

Measurement points	*Ground cover of the plant* (%)
1	85
2	65
3	20
4	20
5	30
6	35
7	100
8	90

DO NOT WRITE IN THIS MARGIN

Marks KU PS

(i) 1 What is the light intensity in the garden where the ground cover of the plant was 100%?

__________ lux **1**

2 What was the percentage ground cover of the plant when the light intensity was 750 lux?

__________ % **1**

(ii) What is the relationship between light intensity and percentage ground cover of the plant?

____

____ **1**

(*c*) Explain how light intensity affects the distribution of the plants in the garden.

____

____ **1**

DO NOT WRITE IN THIS MARGIN

Marks | KU | PS

2. (*a*) The diagram shows part of a food web from a forest.

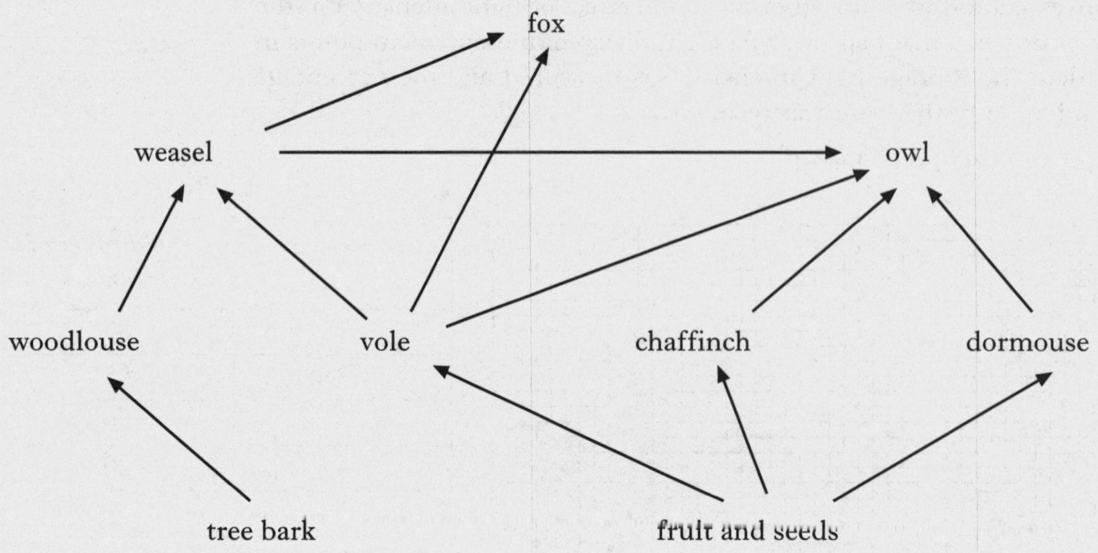

fox

weasel

owl

woodlouse

vole

chaffinch

dormouse

tree bark

fruit and seeds

(i) The numbers of dormice and owls may be affected if the chaffinches were removed from the food web.

1 Underline **one** answer in the brackets and give an explanation.

The dormouse population would { increase / decrease / stay the same }.

Explanation ______________________________

______________________________ **1**

2 Underline **one** answer in the brackets and give an explanation.

The owl population would { increase / decrease / stay the same }.

Explanation ______________________________

______________________________ **1**

(ii) Select a food chain from the web which is made up of four stages.

__________ → __________ → __________ → __________ **1**

DO NOT WRITE IN THIS MARGIN

Marks	KU	PS

2. (continued)

(*b*) A food chain from the ocean is shown below.

plankton → krill → blue whale

Which population in the food chain has the smallest biomass?

______________________________ **1**

[Turn over

DO NOT WRITE IN THIS MARGIN

Marks KU PS

3. (*a*) The grid contains the names of some components of food.

carbon A	hydrogen B	amino acids C
nitrogen D	simple sugar E	glycerol F
fatty acids G	oxygen H	water I

Use letters from the grid to identify the following:

(i) The sub-units of protein molecules ______ **1**

(ii) The sub-units of fat molecules ______ and ______ **1**

(iii) An element found in protein but not in starch ______ **1**

(*b*) Name the structures in the small intestine which provide an increased surface area for absorption.

______________________________ **1**

(*c*) Urea is produced in the liver from the breakdown of digested food molecules. From which food molecules is urea produced?

______________________________ **1**

DO NOT WRITE IN THIS MARGIN

Marks KU PS

4. (*a*) The diagram shows part of the human breathing system.

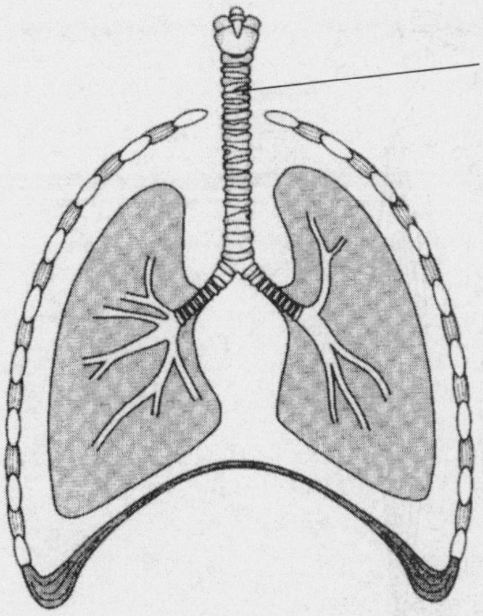

cartilage rings

Describe the function of the cartilage rings.

______________________________ **1**

(*b*) (i) Name the sticky substance that traps inhaled dust particles.

______________ **1**

(ii) Explain how the trapped particles are removed from the breathing system.

____

______________________________ **1**

(*c*) As blood passes through capillary networks in the lungs, oxygen and carbon dioxide are exchanged between the blood and the air sacs.

(i) Describe **one** feature of a capillary network which allows efficient gas exchange.

____

______________________________ **1**

(ii) Name the structures in blood that contain haemoglobin.

______________ **1**

(iii) Explain the function of haemoglobin in the transport of oxygen.

______________________________ **1**

DO NOT WRITE IN THIS MARGIN

Marks KU PS

5. (*a*) The diagram represents phloem tissue from the stem of a plant.

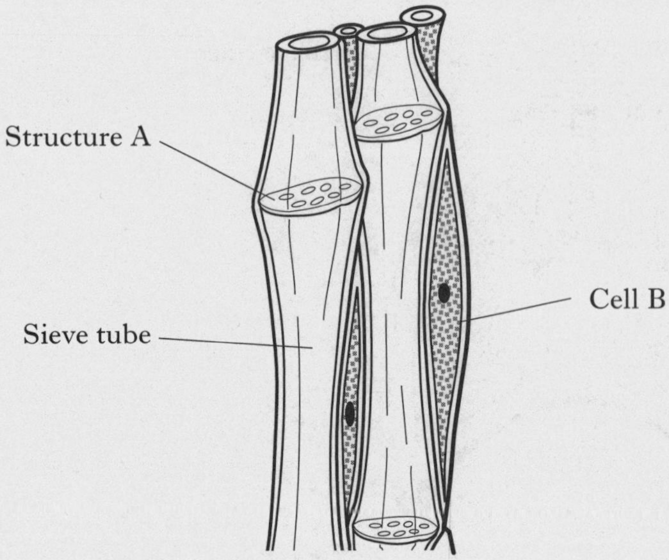

(i) Name Structure A and Cell B.

Structure A ____________________

Cell B ____________________ **2**

(ii) State the function of phloem.

______________________________ **1**

(*b*) (i) Name the leaf tissue where stomata are found.

____________________ **1**

(ii) Name the cells which control the opening and closing of stomata.

____________________ **1**

DO NOT WRITE IN THIS MARGIN

Marks | KU | PS

5. (continued)

(*c*) Leaves were placed in tubes as shown below.

The tubes were left in bright light.

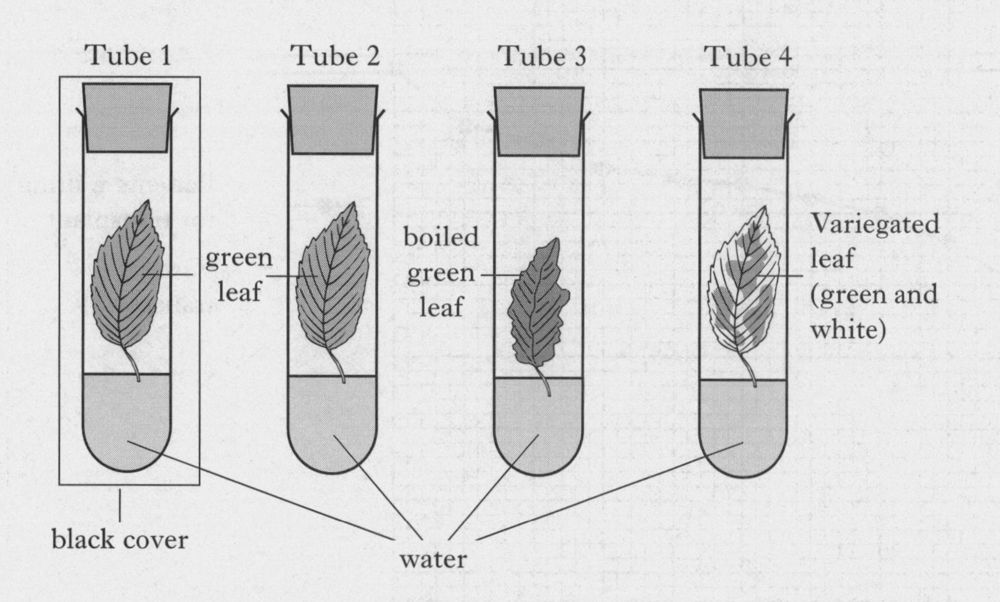

For each of the tubes, tick (✓) the appropriate box in the table to indicate which processes will take place in the leaves.

Process / *Tube*	*Only photosynthesis*	*Only respiration*	*Both*	*Neither*
1				
2				
3				
4				

2

[Turn over

DO NOT WRITE IN THIS MARGIN

Marks KU PS

6. (*a*) The graph shows the number of kidney transplants carried out and the number of patients waiting for a transplant in the UK between 1996 and 2005.

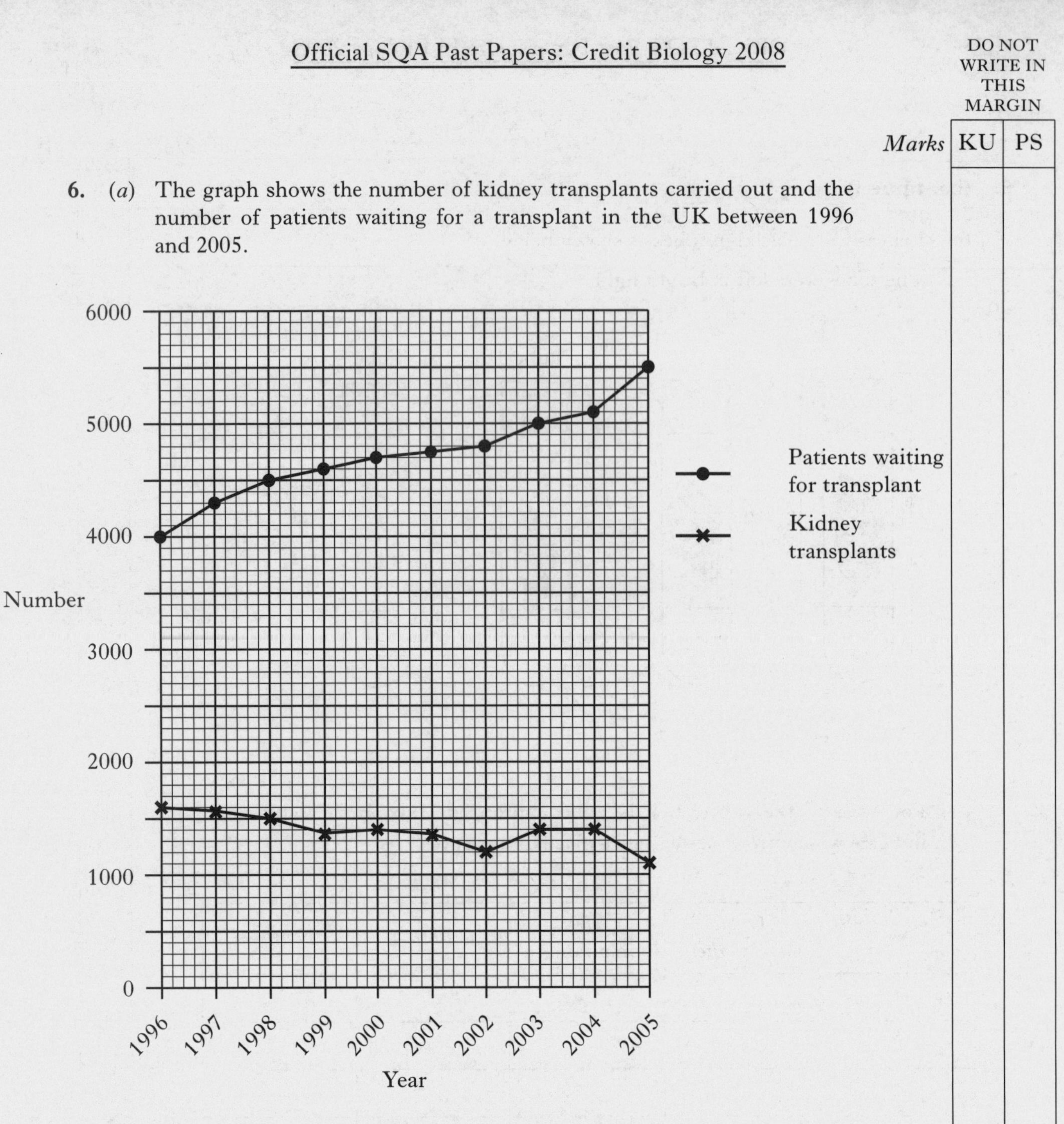

(i) Calculate the average yearly increase in the number of patients waiting for a transplant from 2000 to 2005.

Space for calculation.

Average yearly increase ________ patients per year. **1**

DO NOT WRITE IN THIS MARGIN

6. (*a*) (continued)

Marks	KU	PS

(ii) Calculate the simple whole number ratios of patients waiting for a transplant to the number of kidney transplants carried out for 1996 and for 2005.

Space for calculation.

1996 ______________ : ______________

2005 ______________ : ______________ **1**

patients waiting for a transplant — transplants carried out

(iii) The following statements refer to the data in the graph.

Tick (✓) the box(es) of the correct statement(s).

The number of patients waiting for a transplant increased every year. ☐

The number of transplants carried out decreased every year. ☐

The difference between the number of patients waiting for a transplant and the number of transplants carried out increased every year. ☐ **1**

(*b*) Give **one** advantage and **one** disadvantage of treating kidney failure by transplant compared to treatment using a dialysis (kidney) machine.

Advantage ______________________________

______________________________ **1**

Disadvantage ______________________________

______________________________ **1**

[Turn over

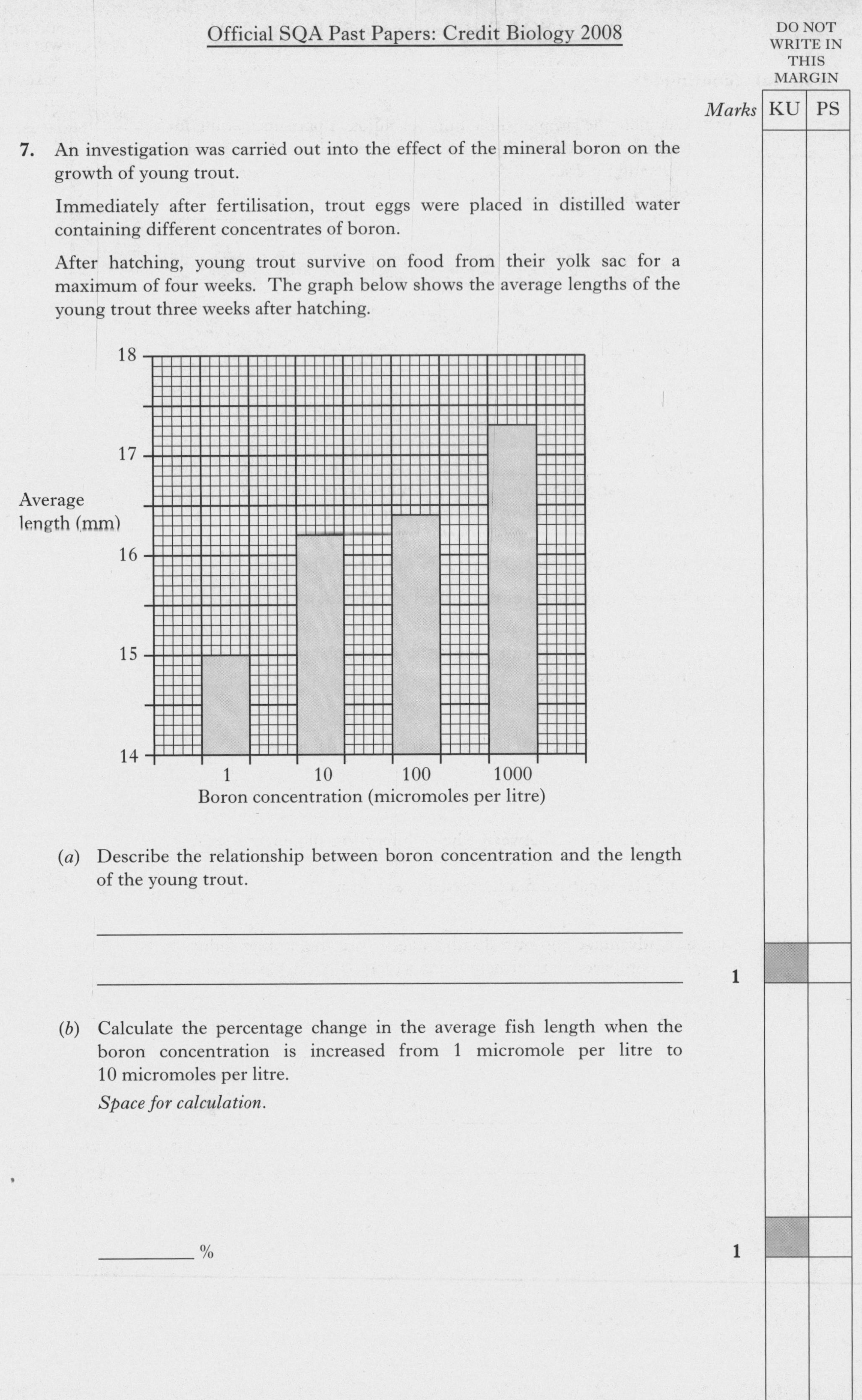

DO NOT WRITE IN THIS MARGIN

Marks | KU | PS

7. An investigation was carried out into the effect of the mineral boron on the growth of young trout.

Immediately after fertilisation, trout eggs were placed in distilled water containing different concentrates of boron.

After hatching, young trout survive on food from their yolk sac for a maximum of four weeks. The graph below shows the average lengths of the young trout three weeks after hatching.

(*a*) Describe the relationship between boron concentration and the length of the young trout.

____

____ **1**

(*b*) Calculate the percentage change in the average fish length when the boron concentration is increased from 1 micromole per litre to 10 micromoles per litre.

Space for calculation.

__________ % **1**

DO NOT WRITE IN THIS MARGIN

Marks	KU	PS

7. **(continued)**

(*c*) Distilled water is the purest form of water available. Give a reason for using distilled water in this investigation.

____

____ **1**

(*d*) Explain why the results would not be valid if the fish were measured more than four weeks after hatching.

____

____ **1**

[Turn over

8. An investigation was carried out into the effect of water concentration on the rate of osmosis.

Details of the apparatus, method used and results are given below.

Apparatus

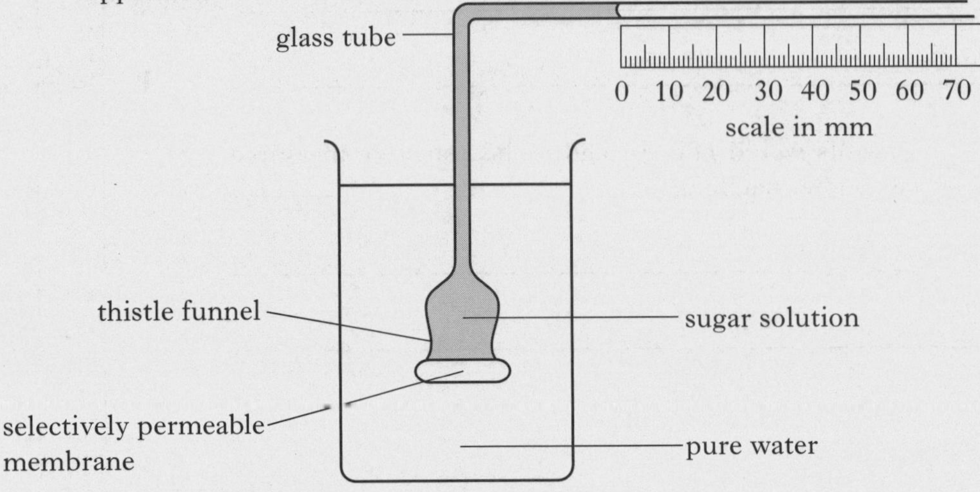

Method

1 A thistle funnel containing 50 cm^3 of 0·5% sugar solution was covered with selectively permeable membrane.
2 The funnel was placed in a beaker of pure water.
3 The scale was positioned with the sugar solution at zero on the scale.
4 The position of the sugar solution was recorded after 30 minutes.
5 The procedure was repeated using 1·0%, 2·0% and 3·0% sugar solutions.

Results

Concentration of sugar solution (%)	*Distance moved by sugar solution in 30 minutes* (mm)
0·5	4·5
1·0	9·0
2·0	18·0
3·0	27·0

DO NOT WRITE IN THIS MARGIN

	Marks	KU	PS

8. (continued)

(*a*) Identify **two** variables not already mentioned that should be kept constant when setting up the investigation.

1 ______________________________

2 ______________________________ **2**

(*b*) Explain the movement of the sugar solution in terms of water concentrations.

____

______________________________ **1**

(*c*) From the results, predict the distance moved by a 3·5% sugar solution in 30 minutes and justify your prediction.

Prediction __________ mm **1**

Justification ______________________________

______________________________ **1**

[Turn over

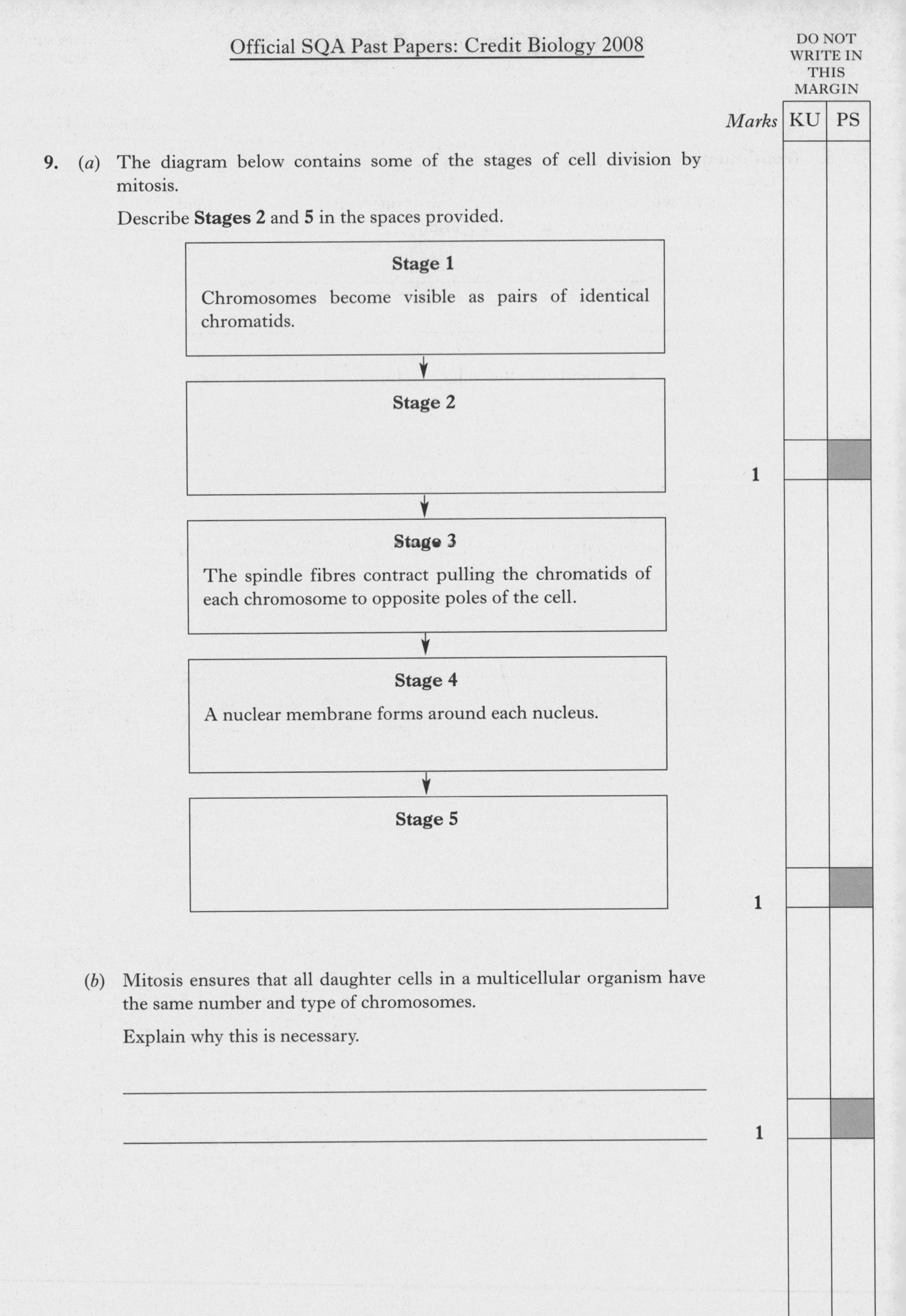

DO NOT WRITE IN THIS MARGIN

Marks KU PS

9. (*a*) The diagram below contains some of the stages of cell division by mitosis.

Describe **Stages 2** and **5** in the spaces provided.

Stage 1

Chromosomes become visible as pairs of identical chromatids.

Stage 2

1

Stage 3

The spindle fibres contract pulling the chromatids of each chromosome to opposite poles of the cell.

Stage 4

A nuclear membrane forms around each nucleus.

Stage 5

1

(*b*) Mitosis ensures that all daughter cells in a multicellular organism have the same number and type of chromosomes.

Explain why this is necessary.

____

____ 1

DO NOT WRITE IN THIS MARGIN

Marks KU PS

10. (*a*) Barley is a plant grown for use in the brewing industry. The photographs below show two varieties of barley that have been produced by selective breeding.

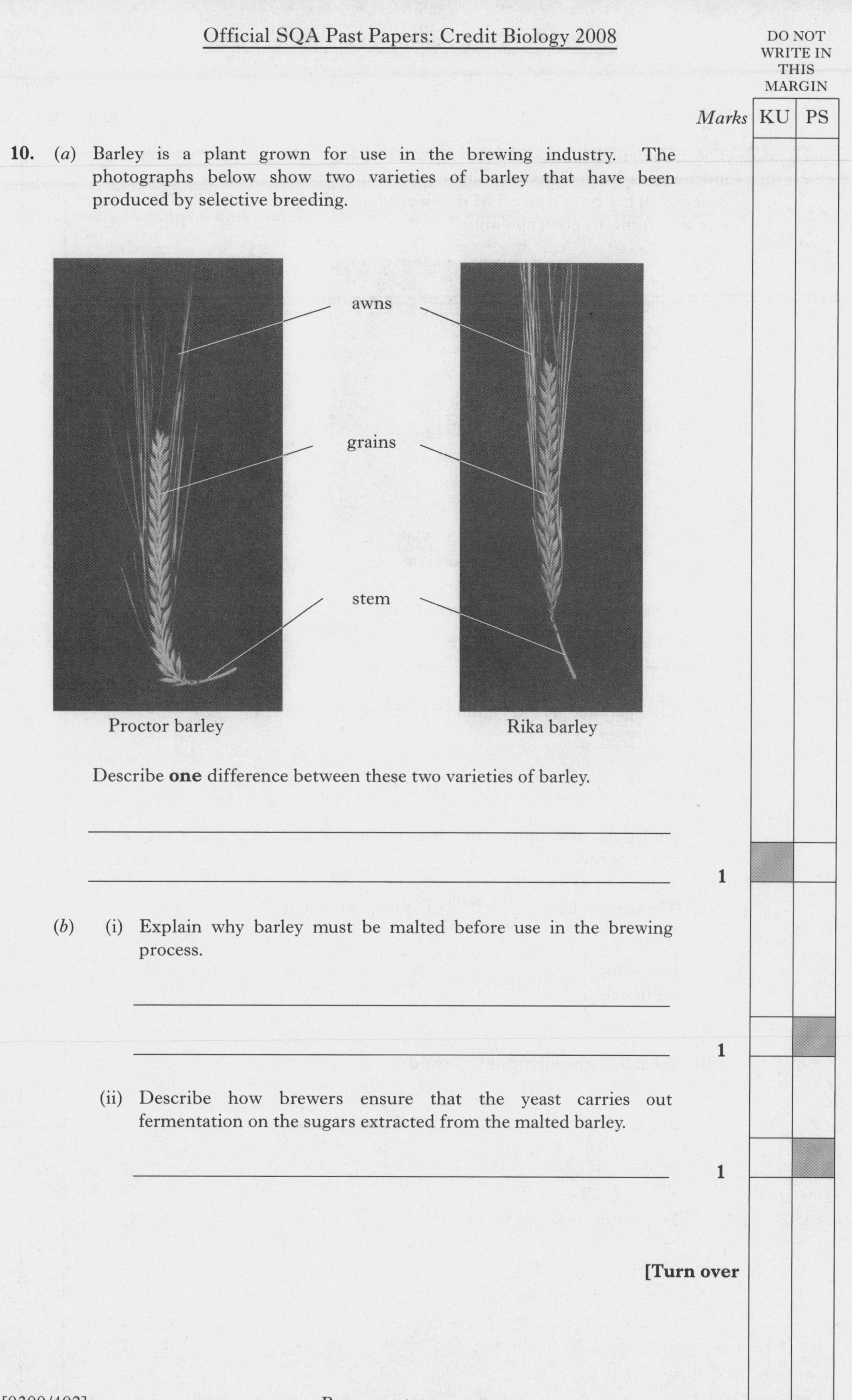

Proctor barley

Rika barley

Describe **one** difference between these two varieties of barley.

____

____ **1**

(*b*) (i) Explain why barley must be malted before use in the brewing process.

____

____ **1**

(ii) Describe how brewers ensure that the yeast carries out fermentation on the sugars extracted from the malted barley.

____ **1**

[Turn over

DO NOT WRITE IN THIS MARGIN

Marks | KU | PS

11. (*a*) The photograph shows a child with dimples. Dimples are small indentations in the cheeks. Their presence is controlled by a single gene which has two forms. The dominant form (**D**) gives dimples. The recessive form (**d**) gives no dimples.

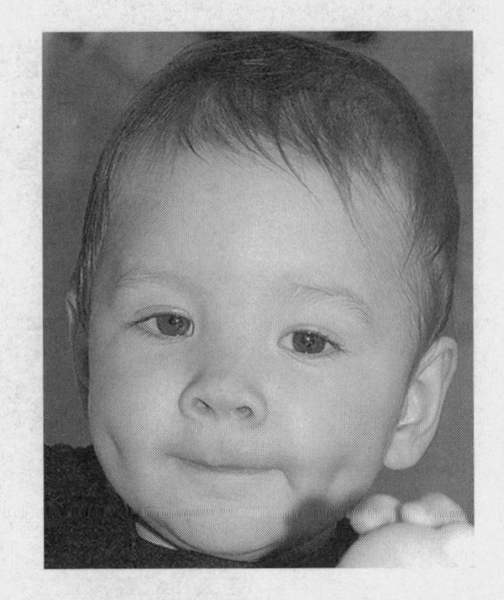

(i) What name is given to different forms of the same gene?

______________________________ **1**

(ii) The parents of the child are known to have the following genotypes.

DD × dd

<u>Underline</u> **one** option in each bracket to make the following sentence correct.

The parents have { the same / different } phenotypes and { the same / different } genotypes. **1**

(iii) What is the genotype of this child?

______________________________ **1**

DO NOT WRITE IN THIS MARGIN

Marks KU PS

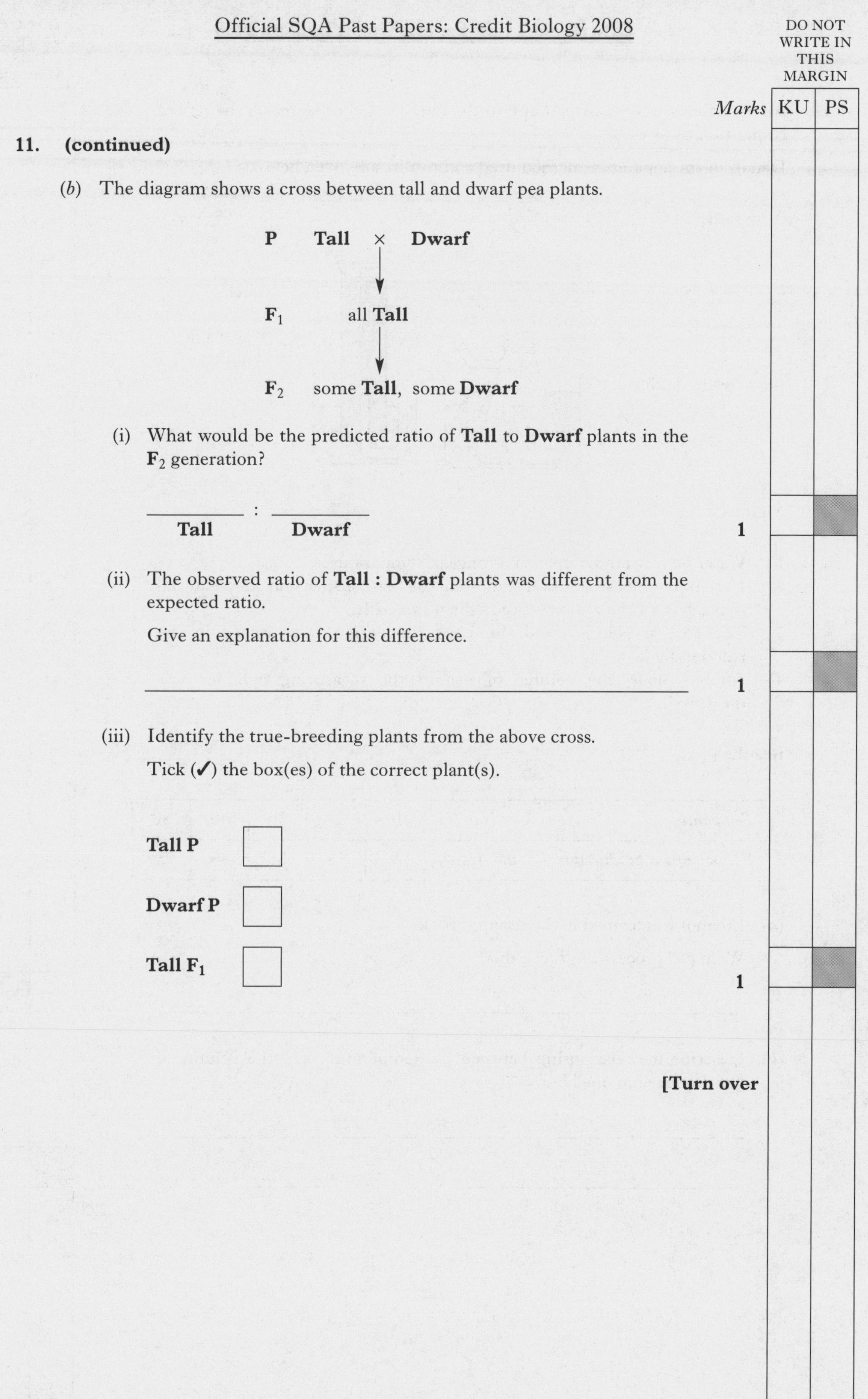

11. (continued)

(*b*) The diagram shows a cross between tall and dwarf pea plants.

(i) What would be the predicted ratio of **Tall** to **Dwarf** plants in the $\mathbf{F_2}$ generation?

______ : ______
Tall **Dwarf** 1

(ii) The observed ratio of **Tall : Dwarf** plants was different from the expected ratio.

Give an explanation for this difference.

______________________________ 1

(iii) Identify the true-breeding plants from the above cross.

Tick (✓) the box(es) of the correct plant(s).

Tall P ☐

Dwarf P ☐

Tall $\mathbf{F_1}$ ☐ 1

[Turn over

DO NOT WRITE IN THIS MARGIN

Marks | KU | PS

12. An investigation was carried out into the effect of temperature on the rate of respiration by yeast.

Details of the apparatus, method used and results are given below.

Apparatus

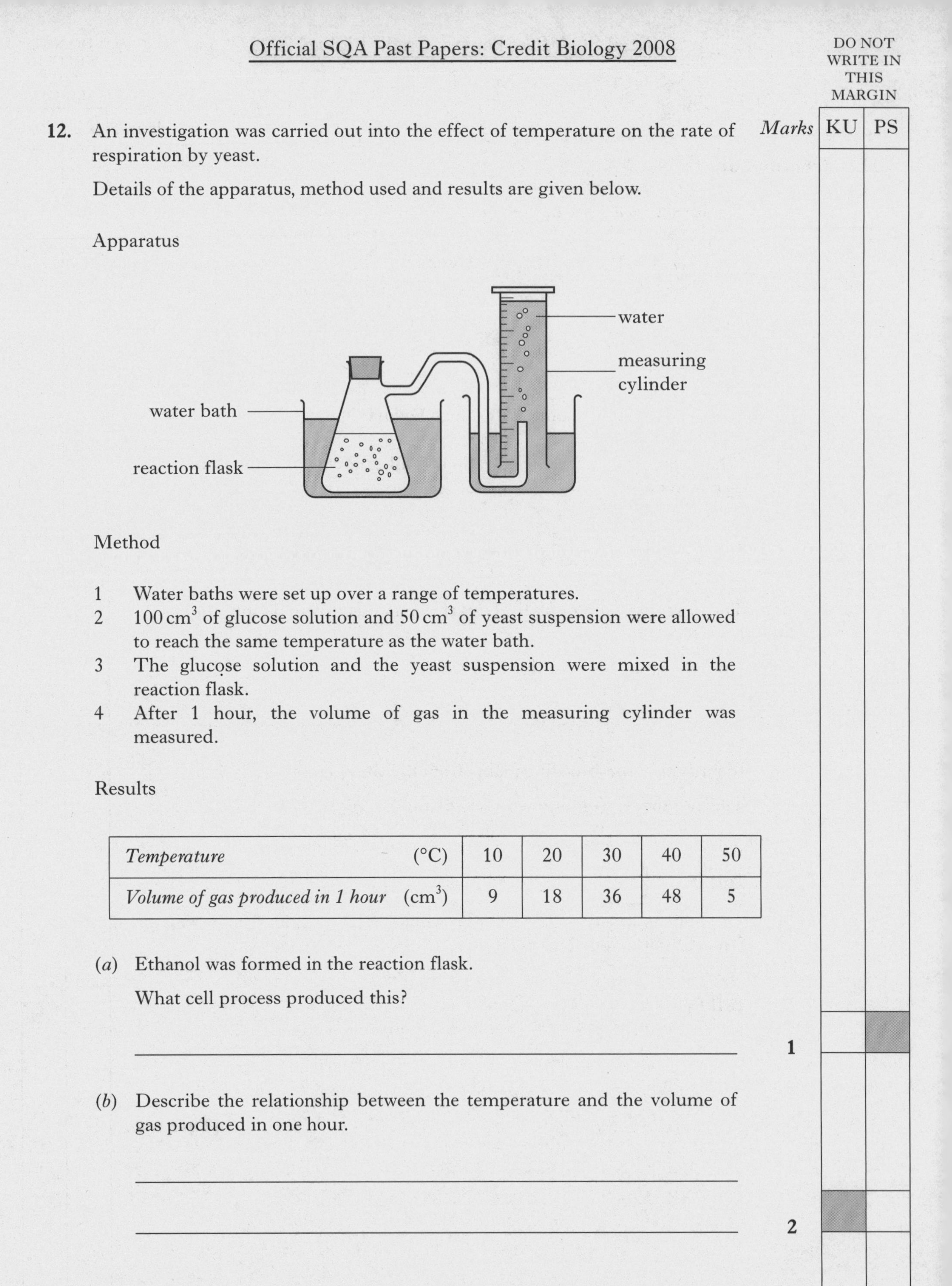

Method

1 Water baths were set up over a range of temperatures.
2 100 cm^3 of glucose solution and 50 cm^3 of yeast suspension were allowed to reach the same temperature as the water bath.
3 The glucose solution and the yeast suspension were mixed in the reaction flask.
4 After 1 hour, the volume of gas in the measuring cylinder was measured.

Results

Temperature (°C)	10	20	30	40	50
Volume of gas produced in 1 hour (cm^3)	9	18	36	48	5

(*a*) Ethanol was formed in the reaction flask.

What cell process produced this?

____ **1**

(*b*) Describe the relationship between the temperature and the volume of gas produced in one hour.

____

____ **2**

DO NOT WRITE IN THIS MARGIN

12. (continued) *Marks* KU PS

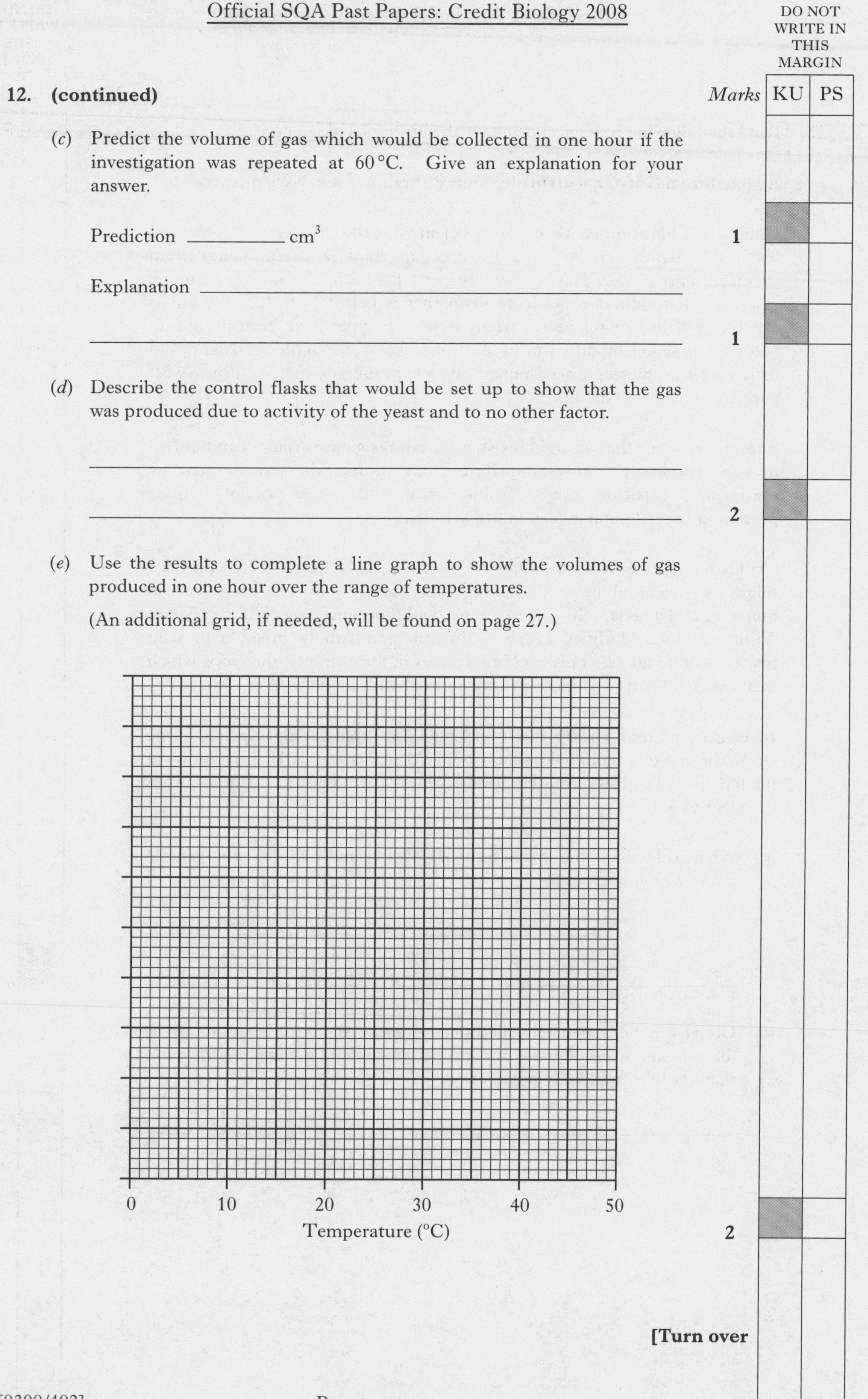

(*c*) Predict the volume of gas which would be collected in one hour if the investigation was repeated at 60 °C. Give an explanation for your answer.

Prediction ____________ cm^3 **1**

Explanation ____

____ **1**

(*d*) Describe the control flasks that would be set up to show that the gas was produced due to activity of the yeast and to no other factor.

____

____ **2**

(*e*) Use the results to complete a line graph to show the volumes of gas produced in one hour over the range of temperatures.

(An additional grid, if needed, will be found on page 27.)

2

[Turn over

DO NOT WRITE IN THIS MARGIN

Marks | KU | PS

13. Read the following passage and answer the questions based on it.

Adapted from ***GM Organisms*** by John Pickrell, www.newscientist.com

Genetic modification (GM) of crops began with the discovery that the soil bacterium *Agrobacterium* could be used to transfer useful genes from unrelated species into plants. The Bt gene is one of the most commonly inserted. It produces a pesticide toxin that is harmless to humans but is capable of killing insect pests. Many new crop types have been produced. Most of these are modified to be pest, disease or weedkiller resistant, and include wheat, maize, oilseed rape, potatoes, peanuts, tomatoes, peas, sweet peppers, lettuce and onions.

Supporters argue that drought resistant or salt resistant varieties can flourish in poor conditions. Insect-repelling crops protect the environment by minimising pesticide use. Golden rice with extra vitamin A or protein-enhanced potatoes can improve nutrition.

Critics fear that GM foods could have unforeseen effects. Toxic proteins might be produced or antibiotic-resistance genes may be transferred to human gut bacteria. Modified crops could become weedkiller resistant "superweeds". Modified crops could also accidentally breed with wild plants or other crops. This could be serious if, for example, the crops which had been modified to produce medicines bred with food crops.

Investigations have shown that accidental gene transfer does occur. One study showed that modified pollen from GM plants was carried by the wind for tens of kilometres. Another study proved that genes have spread from the USA to Mexico.

(*a*) What role does the bacterium *Agrobacterium* play in the genetic modification of crops?

____

____ **1**

(*b*) Crops can be genetically modified to make them resistant to pests, diseases and weedkillers. Give another example of genetic modification that has been applied to potatoes.

____ **1**

DO NOT WRITE IN THIS MARGIN

	Marks	KU	PS

13. (continued)

(*c*) Explain why a plant, which is modified to be weedkiller resistant could be:

(i) useful to farmers.

____ **1**

(ii) a problem for farmers.

____ **1**

(*d*) Give **one** example of a potential threat to health by the use of GM crops.

____ **1**

[Turn over

DO NOT WRITE IN THIS MARGIN

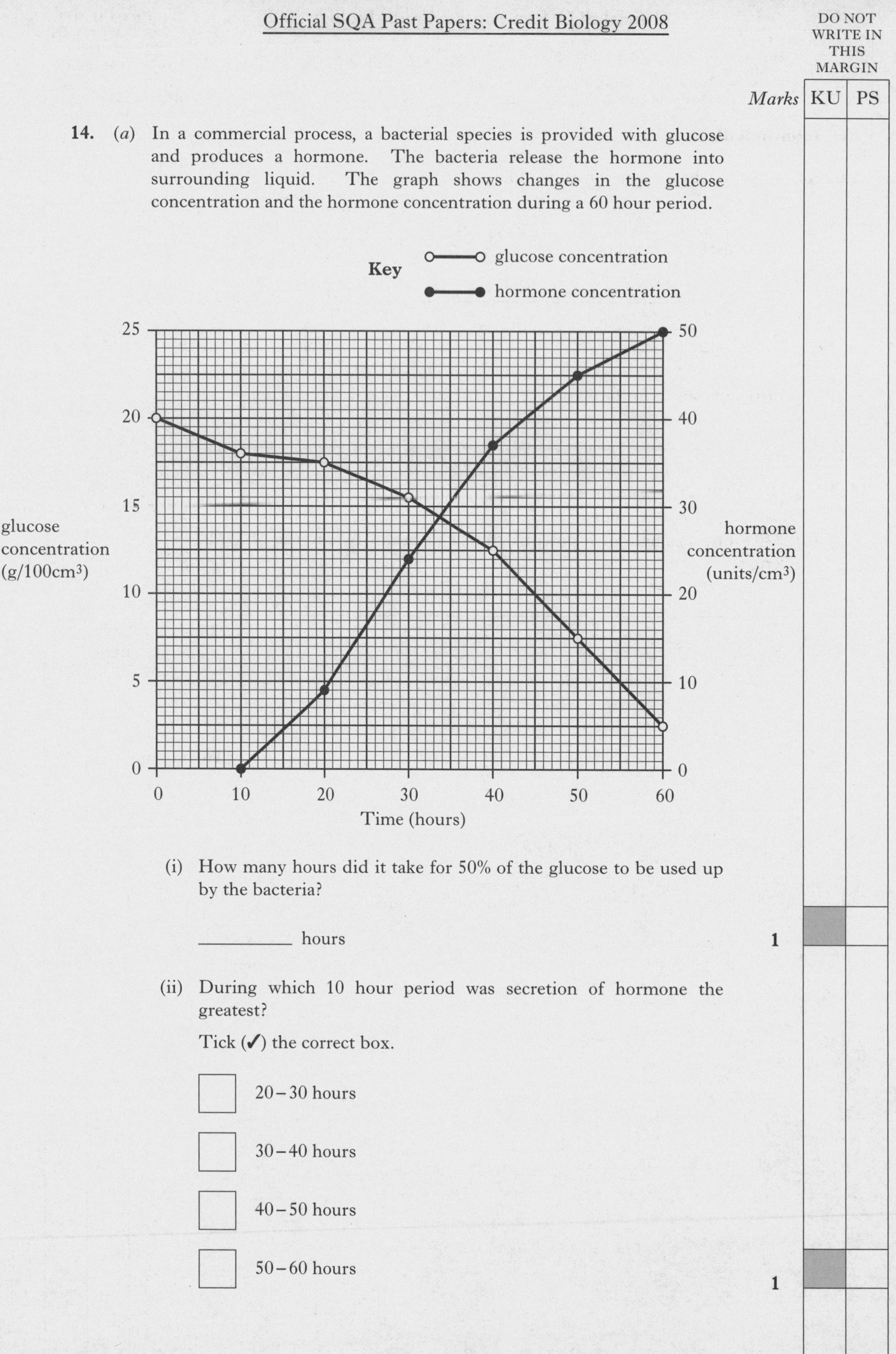

Marks	KU	PS

14. (*a*) In a commercial process, a bacterial species is provided with glucose and produces a hormone. The bacteria release the hormone into surrounding liquid. The graph shows changes in the glucose concentration and the hormone concentration during a 60 hour period.

(i) How many hours did it take for 50% of the glucose to be used up by the bacteria?

__________ hours **1**

(ii) During which 10 hour period was secretion of hormone the greatest?

Tick (✓) the correct box.

☐ 20 – 30 hours

☐ 30 – 40 hours

☐ 40 – 50 hours

☐ 50 – 60 hours **1**

DO NOT WRITE IN THIS MARGIN

Marks | KU | PS

14. (*a*) (continued)

(iii) Calculate the decrease in glucose concentration over the 60 hour period.

Space for calculation.

__________ g/100 cm^3 **1**

(iv) If glucose continues to be used at the same rate as between 50 and 60 hours, predict how many more hours it would be before all the glucose would be used up.

Space for calculation.

__________ hours **1**

(v) During the first 10 hours of the process, energy was being used for functions other than the synthesis of the hormone.

Give **two** pieces of evidence from the graph to support this statement.

1 ______________________________

2 ______________________________ **1**

(*b*) Glucose is a carbohydrate component of food. Which food component contains most energy per gram?

______________________________ **1**

[Turn over for Question 15 on *Page twenty-six*

DO NOT WRITE IN THIS MARGIN

Marks	KU	PS

15. (*a*) In a sewage works, micro-organisms cause the decay of the sewage. What is the benefit to the micro-organisms in carrying out this process?

______________________________ **1**

(*b*) What type of respiration must be carried out by the micro-organisms to ensure complete breakdown of the sewage?

______________ **1**

(*c*) Sewage contains a wide range of materials. What ensures that all these materials are broken down?

______________________________ **1**

(*d*) The table shows the methods of disposal of the sludge obtained from sewage treatment.

Method of disposal of sludge	*Percentage*
Spread on farmland	50
Landfill	10
Dumped at sea	15
Incinerated	20
Other disposal	5

Use the information from the table to complete the pie chart below.

(An additional chart, if needed, will be found on page 27.)

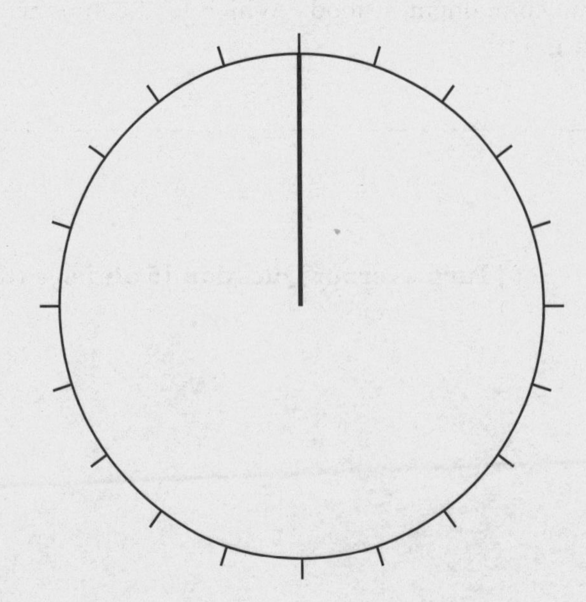

2

[*END OF QUESTION PAPER*]

ADDITIONAL GRAPH PAPER FOR QUESTION 12(*e*)

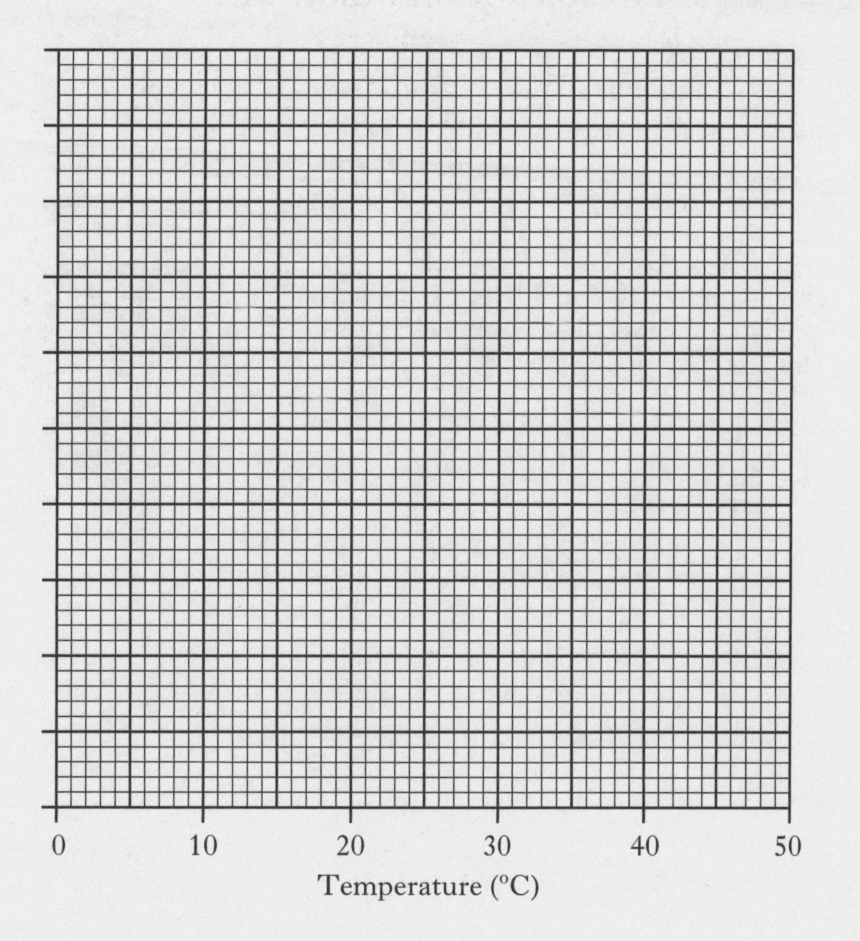

ADDITIONAL PIE CHART FOR QUESTION 15(*d*)

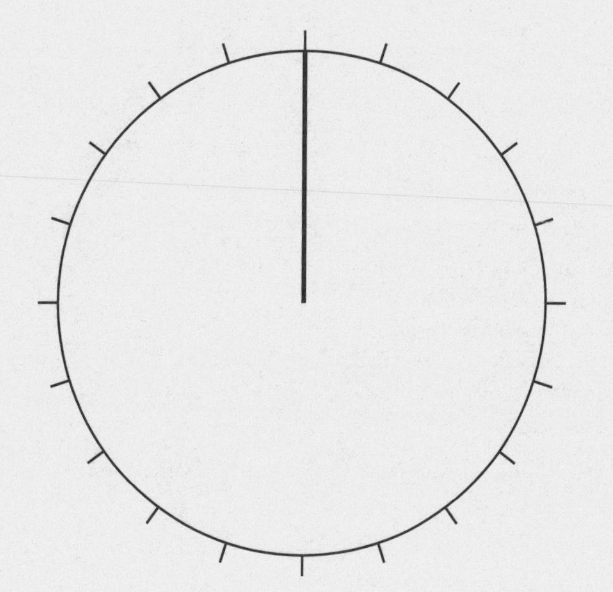

SPACE FOR ANSWERS
AND FOR ROUGH WORKING

[BLANK PAGE]

[BLANK PAGE]

[BLANK PAGE]

[BLANK PAGE]

[BLANK PAGE]

Acknowledgements

Leckie and Leckie is grateful to the copyright holders, as credited, for permission to use their material.
The Institute of Biology for an extract adapted from the article 'Getting a Liking for Lichens' by F. Dobson, taken from Biologist, volume 59 (2006 paper p 16);
Philip Allan Updates for the article 'Invasion of the Chinese Mitten Crab' taken from the Biological Sciences Review Volume 15, Number 2 (2007 paper p 16).

The following companies have very generously given permission to reproduce their copyright material free of charge:
Newsquest Media Group for an extract from The Herald, October 2003(2006 paper 13);
The Newscientist and Reed Business Information Centre for an extract from 'GM Organisms' by John Pickrell(2008 paper p 22).